the history of the earth's crust

THE PRENTICE-HALL FOUNDATIONS OF EARTH SCIENCE SERIES
A. Lee McAlester, Editor

STRUCTURE OF THE EARTH

S. P. Clark, Jr.

EARTH MATERIALS

W. G. Ernst

THE SURFACE OF THE EARTH

A. L. Bloom

EARTH RESOURCES, 2nd ed.

B. J. Skinner

GEOLOGIC TIME, 2nd ed.

D. L. Eicher

ANCIENT ENVIRONMENTS, 2nd ed.

L. F. Laporte

THE HISTORY OF THE EARTH'S CRUST

D. L. Eicher, A. L. McAlester, and M. L. Rottman

THE HISTORY OF LIFE, 2nd ed.

A. L. McAlester

OCEANS, 2nd ed.

K. K. Turekian

MAN AND THE OCEAN

B. J. Skinner and K. K. Turekian

ATMOSPHERES

R. M. Goody and J. C. G. Walker

WEATHER

L. J. Battan

THE SOLAR SYSTEM

J. A. Wood

the history of the earth's crust

DON L. EICHER
University of Colorado

A. LEE McALESTER
Southern Methodist University

MARCIA L. ROTTMAN
Exxon Production Research Co.

PRENTICE-HALL, INC., *Englewood Cliffs, New Jersey 07632*

Library of Congress Cataloging in Publication Data

Eicher, Don L.
   History of the earth's crust.

   Bibliography: p.
   Includes index.
   1. Earth—Crust. I. McAlester, A. Lee (Arcie Lee),
(date). II. Rottman, Marcia L. III. Title.
QE511.E37   1984   551.1'3   83-1103
ISBN 0-13-389999-3
ISBN 0-13-389982-9 (pbk.)

Editorial/production supervision: Maria McKinnon

Interior design and cover design: Lee Cohen

Manufacturing buyer: John Hall

Printed in the United States of America

10 9 8 7 6 5 4 3 2

ISBN 0-13-389999-3
ISBN 0-13-389982-9 {PBK}

Prentice-Hall International, Inc. *London* 637|94339
Prentice-Hall of Australia Pty. Limited, *Sydney*
Editora Prentice-Hall do Brasil, Ltda., *Rio de Janeiro*
Prentice-Hall Canada Inc., *Toronto*
Prentice-Hall of India Private Limited, *New Delhi*
Prentice-Hall of Japan, Inc., *Tokyo*
Prentice-Hall of Southeast Asia Pte. Ltd., *Singapore*
Whitehall Books Limited, *Wellington, New Zealand*

contents

four

the paleozoic earth 73

five

the mesozoic earth 120

six

the cenozoic earth 153

preface

Anyone attempting to summarize several billion years of earth history between the covers of a single volume, even one far larger than this, faces constant decisions about what to leave out. This book is no exception. It emphasizes modern understanding of the changing geography and environments of the earth's *crust*—the outermost skin of rock in which the panorama of earth history is most clearly recorded. The treatment is chronological, beginning with our planet's origin at the birth of the solar system almost 5 billion years ago, and ending with the rise of modern humanity amid the fluctuating glacial climates of the last few hundred thousand years.

Several closely related topics covered in companion volumes of *The Prentice-Hall Foundations of Earth Science Series* are given relatively brief treatment. The principles used in deciphering earth history are summarized in a short introductory chapter; important recent developments are also emphasized at appropriate points throughout the text. More detailed discussions can be found in: *Geologic time*, Eicher and *Ancient Environments*, Laporte. Similarly, the chronologic development of life on earth, a traditional central theme of historical geology, is given only a cursory introduction. A more comprehensive survey can be found in *The History of Life* by McAlester.

D.L.E.
A.L.M.
M.L.R.

the history of the earth's crust

one
unraveling
earth history

SEDIMENTARY ROCKS AND EARTH CHRONOLOGY

Our knowledge of the earth's past rests almost entirely on studies of crustal rocks exposed today at or near the surface. Rocks that make up the modern continents have been slowly accumulating throughout geologic time and their record reaches back nearly 4 billion years. Plutonic igneous and metamorphic belts indicate intervals of mountain-building deformation deep within the crust whereas volcanic and sedimentary rocks reflect changing landscapes at the earth's surface at the time they formed. Of these various kinds of rocks, *sedimentary rocks* are particularly useful because they record normal surface environments and thus provide information on ancient life and climates and earlier states of the ocean and atmosphere. In order to work out the long sequence of events recorded in the many types of crustal rocks, we must first have some means to *correlate* them — that is, to interrelate and fit them together in terms of their relative ages — so that a worldwide chronology of events can be established.

There are numerous ways of establishing the age relations of ancient rocks and they can be grouped into three basic categories: physical correlation, correlation by fossils, and radiometric dating. The first two apply primarily to sedimentary rocks. They are based on the sequential nature of sediment deposition and on progressive changes in the animal and plant remains that the rocks contain. The third method, which applies chiefly to igneous and metamorphic rocks, is based on rates of decay of radioactive atoms incorporated in the rocks when they formed. We shall look briefly at these methods for determining earth chronology and then conclude this chapter with a discussion of the universal time scale of earth history and the evidence for the age of the earth.

Physical Correlation

The physical process of sediment deposition provides a means of establishing relative chronology that is not available for most rocks that solidify from molten magmas. Sediments accumulate under the influence of gravity as relatively thin horizontal layers, called *beds* or *strata*, and they are said to be *stratified*. Each such sedimentary bed is deposited on top of older beds and is, in turn, buried by younger beds. This simple relationship means that when thick sequences of sedimentary rocks are preserved without deformation, the *underlying beds are always older than those overlying them*. (Note that this relationship is not necessarily true of igneous rocks, which are normally intruded from below and may crystallize while overlain by older rocks; see Fig. 1-1.) This seemingly obvious principle—that younger sedimentary rocks always overlie older—is known as the *law of superposition* and it provided the first key to deciphering geologic history when it was recognized late in the seventeenth century.

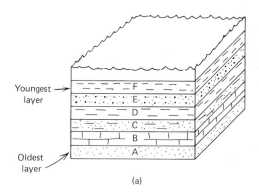

(a)

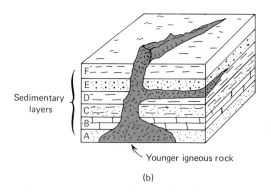

(b)

FIG. 1-1 The law of superposition. (a) In sequences composed entirely of sedimentary rocks, younger beds always overlie older. The sequence shown thus becomes progressively younger from A to E. (b) Igneous rocks, on the other hand, are not necessarily older than overlying rocks. In the sequence shown the igneous unit (G) was intruded into the sedimentary layers (A–F) and thus is younger than the sediments it has intruded, including those that in places overlie it.

Sequences of ancient sedimentary rocks thousands of meters thick, representing many different ages and sedimentary environments, are found on every continent. Once the law of superposition was understood, it became possible to work out the geologic history of a region by observing the progressive upward changes in sediment types. A major difficulty in such studies stems from lack of adequate *exposure* of the rocks. On most parts of the land, surface sedimentary rocks are overlain by a thin veneer of soil and vegetation that largely hides the underlying rock. In such regions the bedrock can generally be observed only in scattered localities, such as where streams have cut into it or where manmade excavations like road cuts, quarries, or mines have exposed it artificially. As a result of these limitations, it is always necessary to correlate the sedimentary rock strata observed at widely separated exposures.

The simplest kinds of sedimentary rock correlations are those based on the physical nature of the rocks themselves. Where exposures are continuous or closely spaced, as in many desert regions, it is possible to follow single beds or groups of beds over long distances (Fig. 1-2). More commonly, physical correlations are made by recognizing similar rock types, or sequences of rock types, in discontinuous exposures (Fig. 1-3).

The difficulty with physical correlations is that they are useful only for rocks originally deposited within limited regions and in the same sedimentary environment or sequence of environments. Physical tracing techniques are use-

FIG. 1-2 In the canyon country of the southwestern United States strata can be traced continuously for long distances. In regions of more vegetation or lower relief outcrops of strata are scattered and their correlation must be inferred chiefly on the basis of physical characteristics.

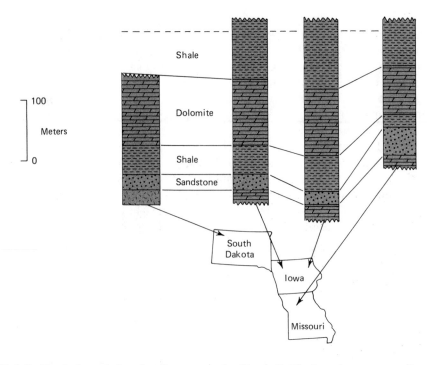

FIG. 1-3 Physical correlation of sedimentary rocks. The similarities in rock sequences often permit interrelation of sedimentary rocks over wide areas. The illustration shows a distinctive sequence of sandstone, shale, dolomite, and another shale that can be recognized over much of the north-central United States.

less between continents or over distances of hundreds of kilometers on a single continent because sedimentary environments and the resulting types of sediment that accumulated changed laterally at any given time in the past just as they do today. Lateral changes within the same stratigraphic interval are called sedimentary *facies* (Fig. 1-4). Facies changes in sedimentary strata may appear

FIG. 1-4 Typical facies changes from nearshore to offshore in a shallow sea. The facies are shown to change by intertonguing.

Calcareous sediments accumulate far offshore beyond the reach of terrigenous sediments

Mud accumulates offshore in deeper water where sand does not reach

Sand deposited in shallow water near shore where longshore currents transport it

Sea
level

Limestone   Shale   Sandstone

as physical differences, such as color, grain size, sorting, mineral content, bedding or other sedimentary structures. Facies changes may also appear as differences in fossil content — that is, in animal or plant remains or traces that become buried in the accumulating sediment and are preserved as part of the rock. Lateral changes in physical characteristics are commonly called *lithofacies* to distinguish them from changes in fossil content or *biofacies*.

Facies changes are instructive for interpreting ancient environments. Continental deposits, such as stream channels and flood plains, tend to change over distances of only a few hundred meters whereas deposits of shallow epicontinental seas may persist essentially unchanged for hundreds of kilometers. Some transitional deposits, such as coal swamps, tidal flats, and lagoons, may be quite local or may persist for long distances. The precise nature of facies changes can be worked out only through accurate correlations, which thus permit local and regional events to be interpreted in detail.

Correlation by Fossils

Physical correlations based on similarities in rock type cannot provide a worldwide chronology, or even a continent-wide chronology, of sedimentary events. What is needed to establish a worldwide chronology is an independent means of determining the relative ages of sedimentary rocks. Fossils provide an approach to such a chronology because the history of life is punctuated by a rapid succession of evolutionary and extinction events. Fossil remains have long provided the most useful and widely applicable means of establishing the relative ages of sedimentary rocks. Around the beginning of the nineteenth century, about a hundred years after the concept of superposition was first appreciated, an English surveyor named William Smith noticed that each unit of sedimentary rock that he encountered in excavations for a canal west of London was characterized by a distinctive assemblage of fossil shells. These characteristic fossils allowed each unit to be easily recognized wherever it occurred. Using this knowledge and the law of superposition, Smith worked out the general sequence of rocks and fossils exposed in England and Wales. The geological map he published in 1815 illustrating his findings was a landmark in our understanding of earth chronology (Fig. 1-5).

Smith's map clearly demonstrated what later came to be known as the "law of faunal succession" (that successive strata can be identified by the fossils they contain). The way was paved for a worldwide geologic time scale, for it was soon discovered that sequences of fossil assemblages very much like those in England occurred in the sedimentary rocks of Europe and even as far away as North America. This discovery showed that widely separated sedimentary rocks could be correlated simply by comparing their fossil remains with those found in other regions where the fossil sequence had been carefully established. As a result of Smith's discovery, a relative time scale of earth history, based on the fossil contents of sedimentary rocks, was established during the first half of the nineteenth century and has been used with only minor modifications ever since.

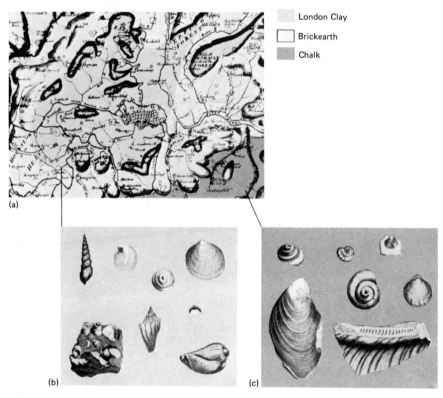

London Clay

Brickearth

Chalk

(a)

(b)

(c)

FIG. 1-5 William Smith's geologic map, published in 1815. (a) Rock units of the London region, (b) Fossil shells in the London Clay, and (c) the Chalk.

Smith and his followers, who established the geologic time scale of earth history, based their work on progressive changes in the types of fossil animals and plants but had little understanding of *why* such changes occurred. We now recognize that the underlying cause of faunal succession is *organic evolution*—the expansions, contractions, and modifications of the biosphere that have taken place since life originated early in earth history. Because of this continuous change of living organisms, the fossil record of past life provides a dynamic framework for judging relative time, a framework that is not available from the physical properties of the sedimentary particles themselves. So successful has the fossil dating technique been that it is still the principal means of unraveling sedimentary crustal history.

The Geologic Time Scale

The last 570 million years of geologic history is known as the Phanerozoic Eon. Within the Phanerozoic, three *eras* bounded by relatively profound, sudden, and worldwide changes in the living organisms preserved as fossils are recognized: the Paleozoic (ancient life) Era, the Mesozoic (middle life) Era, and the Cenozoic (recent life) Era. These three eras are further subdivided into *periods*, each bounded by somewhat less profound changes in the living world. Today 12 periods are recognized. The 7 periods of the Paleozoic Era, listed in order from oldest to most recent, are Cambrian, Ordovician, Silurian, Devonian, Mississippian, Pennsylvanian, and Permian. The 3 periods of the Mesozoic Era are Triassic (oldest), Jurassic, and Cretaceous. The Cenozoic Era includes only 2 periods: Paleogene and Neogene. The 12 geologic periods are, in turn, further subdivided according to their fossil contents into still smaller units called *epochs*.

The oldest sedimentary rocks bearing abundant fossils are those of the Cambrian Period, which begins the Paleozoic Era. Wherever they occur around the world, fossil-bearing Cambrian rocks contain the same distinctive association of shell-bearing marine animals dominated by the long-extinct trilobites, which were distant relatives of modern crabs and shrimp. Cambrian strata everywhere overlie older rocks that lack abundant fossils.

In most regions the rocks below the Cambrian are part of the "basement complex" of older igneous and metamorphic rocks that make up the bulk of the continental crust. In other regions, however, the earliest Cambrian fossils occur above thick sequences of sedimentary rocks that are identical to the overlying Cambrian rocks except that they contain no fossils or only sparse traces of very primitive life that have virtually no stratigraphic value. This abrupt appearance of trilobites, and other relatively advanced kinds of animal life, in the sedimentary record is perhaps the most significant milestone in the earth's long history, for it serves to divide the earth's past into two great divisions: the Phanerozoic Eon above and all of Precambrian time below. Precambrian rocks contain few fossils and therefore provide no worldwide time scale based on changing life. "Phanerozoic (exposed life) time" and the worldwide applicability of a time scale based on fossils begin with the great Cambrian expansion of animal life.

Even in Phanerozoic rocks a chronology of earth history based only on fossils has limitations. In the first place, not all sedimentary rocks contain fossils. In general, they are most common in marine rocks, particularly those that accumulated on the shallow, submerged continents rather than in deep ocean basins. Bottom-dwelling life, particularly shell-bearing life, is abundant today almost everywhere on the shallow ocean floor and the same was also true for much of the geologic past. In contrast, most terrestrial sediments contain very few fossils because even hard skeletal remains decay, weather, and disappear quickly when exposed to the atmosphere. Only in relatively rare terrestrial sediments deposited beneath the standing waters of lakes or swamps are animal and plant remains common.

Because fossils are most abundant in sedimentary rocks of marine origin, it follows that it is usually far easier to establish the age relations of such rocks than it is for those deposited in terrestrial environments. If sea level had always stood as low in relation to the continents as it does today, there would be few ancient marine sediments preserved above sea level and correlations of sedimentary rocks on continents would be correspondingly difficult. Fortunately, however, during portions of the geologic past shallow seas covered much of the surface of the continents. These times are represented by fossil-rich marine sedimentary rocks that cover a large fraction of the continental surfaces, even in regions that today lie far above sea level. Other portions of the geologic

FIG. 1-6 (a) Angular unconformity: The steeply dipping strata are Miocene sandstones and shales that were tilted and planed off by erosion prior to the deposition of the overlying Pleistocene conglomerate beds, Socorro County, New Mexico. Horse and rider give scale. (U.S. Geological Survey) (b) Disconformity: The parallel relationship between the Mississippian Redwall Limestone above and the Cambrian Muav Limestone below in the Grand Canyon, Arizona, suggests that there was no significant tectonic activity and little erosion during the approximately 140 million years that is represented at the contact between them. Vertical relief in the photo is about 1000 meters.

past, during which most continental areas stood above sea level as they do today, are largely unrepresented in the sedimentary record. Thus the continents' veneer of sediments never reveals the *complete* sequence of surface events but only those times and places where sediment accumulation was taking place.

Even areas that receive sediment during a given time may, because of mountain building, lowered sea level, or other factors, be subject to later erosion that removes their record of earlier events. For these reasons, nowhere do sedimentary rocks reflect a continuous sequence of local sediment accumulation through a very long interval of earth history; instead the sedimentary veneer at any one place always shows discontinuities representing long time intervals during which either no sediment accumulated in the area or previously deposited sediment was eroded away. Such discontinuities are called *unconformities*. Even though unconformities reflect intervals of time that are not directly recorded by sedimentary rocks, it is often possible to make inferences about the events that took place during that time from the nature of the contact between the rocks above and below the surface of unconformity. Some circumstances where it is possible to do so are illustrated in Fig. 1-6.

Still other difficulties arise in applying the geologic time scale based on fossils. Fossils, of course, are never found in igneous rocks and only rarely can they be recognized in metamorphosed sedimentary rocks. In many cases, igneous rocks may be interbedded with, or otherwise related to, fossil-bearing sedimentary rocks in such a way that their relative ages are apparent (Fig. 1-7). It is difficult or impossible, however, to relate most igneous and metamorphic rocks to the ages established by fossils in sedimentary rocks.

A final limitation of the time scale based on fossils is that it provides only

FIG. 1-7 Typical relationships of igneous and fossil-bearing sedimentary rocks. (a) The age of the igneous intrusion B is bracketed by the ages of sedimentary unit A, which is clearly older than the intrusion, and sedimentary unit C, which is clearly younger because it was deposited on a surface of erosion that truncated both the intrusive and sedimentary sequence A. (b) Each lava flow, having formed at the earth's surface, is younger than the sedimentary rock unit below it and older than the sedimentary rock unit that overlies it.

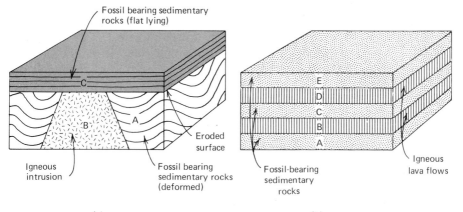

(a)   (b)

relative, not absolute, ages. Fossil-bearing rocks can be readily placed in their correct order, based on the continuous change of the living world; yet fossils provide no means of knowing how long these changes required *in years*. A particular evolutionary sequence might have occurred over thousands, millions, or even hundreds of millions of years, but the fossils themselves provide no direct means for determining this period of time.

NUCLEAR CLOCKS AND EARTH CHRONOLOGY

The time scale of earth history based on fossil remains has been established for well over a hundred years; for most of this time, fossils provided the *only* worldwide scheme for determining earth chronology. Within about the last 25 years, however, fossil dating techniques have been supplemented by powerful new chronologic tools based on the radioactive decay of certain kinds of atoms.

Radioactive Decay

Each type of atom is distinguished from all others by the number of protons and neutrons in its nucleus. The number of protons determines the *element*; the number of protons and neutrons together gives the atom its mass. Thus the proton number and mass number together specify a single, particular kind of atom. For "particular kind of atom" we can use the word *nuclide*. Two different nuclides that contain the same number of protons but different numbers of neutrons belong to the same element but have different masses. Such nuclides are referred to as *isotopes* of that element. Two common isotopes of uranium (proton number 92), for example, are uranium-235 and uranium-238. Most naturally occurring nuclides are stable and have no tendency to decay to other nuclides. A few types, however, are unstable and change spontaneously to a lower energy state by *radioactive decay*. During decay the radioactive nuclide, which is called the "parent," changes into another kind of nuclide called the "daughter." The rate of change is constant and this fact makes possible the use of radioactive decay as a tool of earth chronology.

All radioactive nuclides that occur in nature come from one of two sources. The *primary radioactive nuclides* have such extremely slow rates of decay that they have persisted since the earth first formed. About 20 such nuclides have been detected. Four are widespread and abundant enough to be generally useful as chronologic tools: potassium-40, which decays to argon-40; rubidium-87, which decays to strontium-87; uranium-235, which decays, through a series of intermediate radioactive nuclides, to lead-207; and, finally, uranium-238, which decays, through an intermediate series of nuclides, to lead-206.

In addition to these long-lived radioactive nuclides left over from the earth's formation, a second group of much shorter-lived radioactive nuclides is continually being produced either by cosmic rays in the earth's upper atmosphere or as intermediate products of uranium and thorium decay. Some of these short-lived radioactive nuclides have found special use in dating materials produced during the last few tens of thousands of years. Carbon-14 is the most widely used of this group.

Radiometric Dating

Each radioactive nuclide decays to a daughter nuclide at its own unique rate, which is a constant, unchanging property of that nuclide. Radioactive decay occurs entirely in the atomic nucleus and the rate of decay is independent of external conditions, such as heat or pressure. The rate is even unaltered by chemical changes, such as oxidation or reduction of the parent atom, because such changes involve only the orbital electrons and not the nucleus. So if a radioactive nuclide is incorporated into a mineral or rock when it crystallizes, the proportion that decays to the radiogenic daughter nuclide is a function only of the elapsed time since the crystallization event.

In the process of radioactive decay the nucleus of a parent radioactive atom changes in one of three ways (Fig. 1-8). It emits an alpha particle, it emits

FIG. 1-8 Examples of the three principal processes of radioactive decay. The examples shown are long-lived nuclides used for radiometric dating.

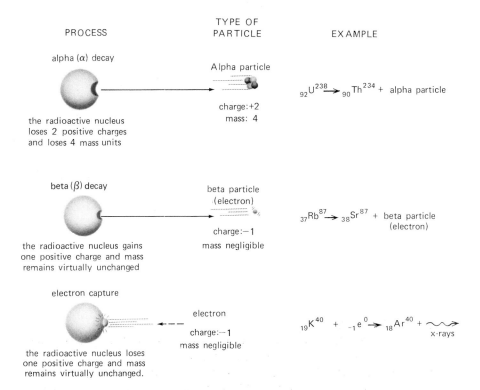

PROCESS — TYPE OF PARTICLE — EXAMPLE

alpha (α) decay

Alpha particle

charge: +2
mass: 4

the radioactive nucleus loses 2 positive charges and loses 4 mass units

$$_{92}U^{238} \longrightarrow {}_{90}Th^{234} + \text{alpha particle}$$

beta (β) decay

beta particle (electron)

charge: −1
mass negligible

the radioactive nucleus gains one positive charge and mass remains virtually unchanged

$$_{37}Rb^{87} \longrightarrow {}_{38}Sr^{87} + \text{beta particle (electron)}$$

electron capture

electron

charge: −1
mass negligible

the radioactive nucleus loses one positive charge and mass remains virtually unchanged.

$$_{19}K^{40} + {}_{-1}e^{0} \longrightarrow {}_{18}Ar^{40} + \text{x-rays}$$

a beta particle, or it captures an electron. In alpha decay the nucleus of the parent atom loses 2 protons and 2 neutrons; the mass number thus decreases by 4 and the proton number decreases by 2. In beta decay the nucleus emits a high-energy electron and one of its neutrons turns into a proton; the proton number increases by one but mass remains unchanged. In electron capture a proton in the nucleus picks up an orbital electron and turns into a neutron, thus decreasing the proton number by one. Again, mass remains unchanged.

The principle of radiometric dating is comparable to an hourglass. Turn the glass over and sand runs from the top chamber to the bottom. As long as some sand remains in the top, the amount in the top relative to the amount accumulated in the bottom provides a measure of the time that has elapsed. The sand in the top of the hourglass represents decaying radioactive parent atoms and that in the bottom accumulating daughter atoms. Just as the hourglass must be sealed so that sand cannot escape through the sides, so the atomic lattice structure of the mineral must be able to hold both the parent and the daughter atoms without allowing any to escape or, for that matter, to enter from an external source. In other words, the system must be closed.

Unlike the passage of sand through an hourglass or other *linear* rates of depletion, radioactive decay occurs at a *geometric* rate (Fig. 1-9). Each individual atom of a given radioactive isotope has the same probability of decaying within the next year and this probability remains the same no matter how long the material being dated has been in existence. The probability of decay is expressed by a number called the *decay constant*, λ, which simply stipulates the *proportion* of atoms of that particular nuclide that always decays in a year. The actual *number* of atoms that will decay is λN, where N is the number of radioactive parent atoms present in the system at the beginning of the year. At the beginning of the next year the number of radioactive parent atoms is, of course, smaller, having decreased by λN. Thus the actual number of atoms to decay the second year is smaller and the number decreases with each successive year. The total time required for *all* radioactive atoms in a given system to decay cannot be specified. In theory it is infinite. It is a simple matter, how-

FIG. 1-9 Changing ratios of parent and daughter atoms as a result of radioactive decay. With the passage of each half-life period, half of the parent atoms that existed at the beginning of the period decay. As the parent atoms decay, they are replaced by daughter atoms.

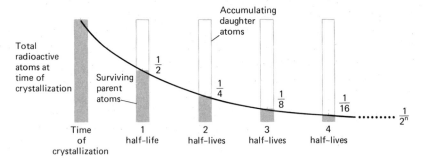

ever, to specify the time required for half the atoms of a particular radioactive nuclide to decay. This time period is called the *half-life*. The end of one half-life period marks the beginning of another. So if a quantity of a radioactive nuclide is segregated, half of the initial number of atoms ($N/2$) remain after one half-life period, and half of them or one-fourth ($N/4$) remain at the end of the next half-life period, and half of those or just one-eighth ($N/8$) remain at the end of the next half-life period, and so on. The number of surviving parent atoms at the end of many (n) half-lives is thus $N/2^n$. This simple relationship is the basis of all radioactive clocks (Fig. 1-9). Each radioactive nuclide has its own half-life; some have a duration of microseconds, others, trillions of years. A given nuclide is best suited to measure ages that are about the same order of magnitude as its half-life.

When rocks containing radioactive atoms first crystallize, ideally they contain no atoms of the radiogenic daughter. The initial daughter–parent ratio is zero and therefore the indicated age is zero. With time, the decay of radioactive parent atoms produces radiogenic daughter atoms within individual mineral grains that make up the rock. Knowing the decay constant of the radioactive parent, we need only measure the ratio of radiogenic daughter and parent nuclides in these mineral grains in order to calculate the *radiometric age* in years before the present. Figure 1-9 shows graphically that with each increment of time the parent–daughter ratio is unique. The common radiometric age determination methods are summarized in Table 1-1.

Because potassium is a constituent of common rock-forming minerals, such as mica, feldspar, and hornblende, the potassium–argon method has proven valuable in dating many types of rocks. Igneous extrusives and other rocks that have never been buried deeply generally give reliable potassium–argon ages. Above temperatures normally attained at a depth of around 5 kilometers, however, some of the argon gas escapes from crystals in which it is produced. Igneous and metamorphic rocks do not begin to retain all their argon until they have cooled to 200°C or so. If they crystallize at great depth, this temperature may not be reached until several million years after crystallization. This tendency for argon loss is the chief disadvantage of the potassium–argon method.

With a half-life of 1300 million years, potassium–argon dating is applicable to the oldest Precambrian rocks. Tiny amounts of argon can be measured with great precision and the potassium–argon whole rock method has also been applied successfully to volcanic rocks as young as 100,000 years.

Rubidium is not a primary component of any common minerals, but it does occur in trace amounts in mica, feldspar, pyroxene, amphibole, and olivine, all of which are common and all of which may be used for rubidium–strontium age determinations. The rubidium–strontium method is useful for both igneous and metamorphic rocks and is perhaps the most valuable technique available for dating metamorphic rocks. Strontium becomes stabilized in minerals much more quickly, and radiogenic strontium-87 begins to ac-

Table 1-1 The Most Important Methods of Radiometric Age Determination

Parent Nuclide	Half-life (years)	Daughter Nuclide	Materials Commonly Dated
URANIUM-238	4500 million	Lead-206	Zircon Uraninite Pitchblende
URANIUM-235	700 million	Lead-207	Zircon Uraninite Pitchblende
POTASSIUM-40	1300 million	Argon-40	Muscovite Biotite Hornblende Glauconite Sanidine Whole volcanic rock
RUBIDIUM-87	47,000 million	Strontium-87	Muscovite Biotite Lepidolite Microcline Glauconite Whole metamorphic rock
CARBON-14	5730		Wood, peat, bone, marine shells

cumulate much earlier and at much higher temperatures, than radiogenic argon-40. Thus for metamorphic rocks the rubidium–strontium method generally yields better results than any other. Its chief disadvantage is that rubidium-87 has a very long half-life; consequently, comparatively young geologic events cannot be measured very accurately.

Both uranium-238 and uranium-235 are radioactive. Minerals rich in uranium are rare and this is the chief drawback to the uranium–lead method. Some widespread minerals, however, contain uranium in trace quantities (as impurities). The most common is the igneous mineral zircon ($ZrSiO_4$), which typically contains about one part per thousand ($^0/_{00}$) uranium. Small quantities of zircon occur in granitic rocks of many ages and so uranium–lead dating has far more utility than it would if it depended on uranium minerals alone.

When a sample is analyzed for uranium and lead and the calculations are completed, the uranium-235/lead-207 and uranium-238/lead-206 ages should agree, provided that the mineral has remained a closed system. This provides a convenient cross check on the results of the age determination. If the ages do not agree, the system has not remained closed and the ages are both younger than the sample's actual age.

In all three dating methods mentioned the ratios of parent to daughter isotopes are measured to determine the age of the rocks. The carbon-14 method uses a slightly different principle. Carbon-14 is continually created in the upper atmosphere by cosmic ray bombardment of nitrogen-14 (Fig. 1-10). The newly created radioactive carbon is quickly oxidized to carbon dioxide, which is taken up by growing plants.

Carbon-14 thus makes up a tiny proportion of all carbon atoms in newly created organic matter. When a plant dies, it ceases to fix atmospheric CO_2 and in time the amount of carbon-14 progressively decreases as the radioactive carbon decays back to nitrogen-14. Thus the ratio of carbon-14 to the non-radioactive carbon in the plant provides a measure of the time that has elapsed since the plant died. The shells and bones of animals that eat plants or extract calcium carbonate from the oceans have also been dated successfully by this method.

The carbon-14 method depends on the special assumptions that (1) the rate of carbon-14 production in the upper atmosphere is nearly constant and (2) the rate of assimilation of carbon-14 into living organisms is rapid relative to the rate of decay. These assumptions seem valid. Radiocarbon dating is useful for only the last brief portion of geologic time because its half-life is very short (5730 years), but many significant events have occurred within this short time span, including worldwide changes in climate, sea level, and coastal sedimentation processes. The method has also been of great use to archeologists.

FIG. 1-10 The carbon-14 cycle. Radioactive carbon-14 is produced from atmospheric nitrogen by cosmic rays (inset). It then enters into carbon dioxide molecules and becomes incorporated into growing plants and thence into animals. When the plant or animal containing the carbon-14 dies, no new carbon-14 is incorporated into its tissues or skeleton. The amount of carbon-14 decreases as radioactive decay proceeds and the amount of carbon-14 remaining is used to date the wood, bone, or shell fragment.

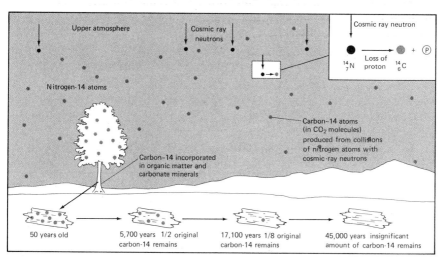

An Absolute Time Scale

Perhaps the most significant achievement of radiometric dating over the past two decades has been to provide an absolute calibration, in years, for the traditional sedimentary time scale. Since the nineteenth century attempts had been made to estimate the actual duration of the Paleozoic, Mesozoic, and Cenozoic eras, without much attention being given to the Precambrian. Most such attempts were based on rates at which sediments are accumulating today. When these rates were extrapolated into the geologic past and applied to the stratigraphic record, they suggested that the geologic eras must have required many millions of years each, but only with the advent of radiometric dating could actual durations be determined.

Radiometric dating techniques are usually applicable only to the minerals of igneous or metamorphic rocks; for this reason, radiometric dating of the sedimentary time scale requires rather unusual associations of sedimentary and igneous rocks (Fig. 1-4). Such associations are far from common. Fortunately, however, enough are known to give a reasonably complete calibration, which is shown on the time scale at the back of this book. The early Cambrian expansion of animal life took place about 570 million years ago. The oldest Precambrian rocks so far dated are about 4000 million years old. Thus the Phanerozoic time scale, based on fossils, represents only the latest 15 percent of the time since the formation of the oldest known crustal rocks. Within this 570 million years of Phanerozoic time, the Paleozoic Era accounts for about the first 345 million years, the Mesozoic Era the next 160 million years, and the Cenozoic Era only the last 65 million years. The average length of the 12 Phanerozoic periods is about 50 million years, but radiometric dating has shown that the changes in the living world that bound the periods were not regularly spaced. Instead the periods range in age from a minimum of about 20 million years to a maximum of about 70 million years. Radiometric dating has also shown that the great ice sheet expansions and contractions of Pleistocene time—events that had such a profound effect on the earth's present climates, landscapes, and sediments—have been concentrated in about the last 2 million years, less than 1 percent of the total of Phanerozoic time.

In addition to furnishing an absolute calibration for the sedimentary time scale, radiometric dating has provided the *only* chronologic tool applicable to the great mass of Precambrian rocks, which represent over 3400 million years of earth history. Before the widespread application of radiometric dating, no worldwide chronology was available for Precambrian rocks. As with younger, fossil-bearing rocks before the discovery of the law of faunal succession, local sequences of Precambrian rocks, and the events they record, had been worked out by purely physical means, but no reliable ways existed to interrelate these events over long distances. Radiometric dating techniques have now revolutionized our understanding of the enormous span of Precambrian history.

THE AGE OF THE EARTH

The oldest crustal rocks so far dated—granites from southwestern Greenland—have radiometric ages of about 4000 million (or 4 billion) years. In other regions metamorphosed sedimentary rocks surround similar ancient granites that were injected into, and thus are younger than, the deformed sediments that surround them. Clearly, then, crustal erosion, sediment deposition, and the formation of sedimentary rocks were all taking place very early in earth history. Apparently these processes have completely recycled and thus obliterated any original rocks formed as the crust first consolidated. Direct dating of crustal rocks can therefore only indicate that the earth is older than 4000 million years. How much older? For the answer, we must turn to less direct evidence of two types.

The first type comes from radiometric dating of meteorites and rocks recovered from the surface of the moon. All the solid material of the solar system, including the earth, moon, and small particles that fall on the earth as meteorites, are believed to have had a common origin in the solar nebula. Neither the rocks of the moon nor presumably meteorites have been subjected to the intense, continuous erosion and weathering suffered by rocks of the earth's crust. Consequently, some might be expected to give ages indicating the time that they and, by extrapolation, the earth first consolidated from solar matter. Most significantly, almost all meteorites have radiometric ages between 4500 and 4700 million years, suggesting that the earth has about the same overall age. The oldest rocks so far recovered from the surface of the moon agree remarkably well. They have ages of about 4600 million years, which suggests that the earth is at least that old.

A second type of indirect evidence for the overall age of the earth also exists; it is based on the present-day abundance of the various nuclides of lead that occur in minerals of the earth's crust. Natural lead is a mixture of four stable nuclides: lead-204, -206, -207, and -208. Three of them (206, 207, and 208) are produced by the radioactive decay of uranium and the less common radioactive elements. The fourth, lead-204, is not produced by radioactive decay. All the lead-204 that is present on the earth today originated when the earth was formed whereas only a part of present-day lead-206, -207, and -208 originated at that time. The rest has been slowly added through the course of earth history by radioactive decay of other elements. If there were some means of determining the *original* relative abundance of the four nuclides at the time the earth was formed, then the earth's age could be estimated by first measuring the *present* relative abundance and then calculating the time required for the additional lead-206, -207, and -208 to have been added by radioactive decay (the decay rates producing each are constant and are precisely known from laboratory measurements). Although there is no direct way to estimate the original lead abundances from crustal rocks, certain lead-bearing meteorites that contain no uranium or other radioactive elements that decay into lead are thought to provide a reasonable approximation of the earth's original lead. Using this information, about 4600 million years of radioactive decay would be necessary to produce lead having the average nuclide abundances found on earth today. This estimate further confirms the suggestion that the earth originated about 4600 million years ago.

two

beginnings

The very elements that make up the earth were produced in the interior of stars and were later assembled together in interstellar space to form the planets. As we have seen, the actual accumulation of the earth took place about 4.6 billion years ago. But precisely how did the earth accumulate? Was it ever a liquid? Have the core, mantle, and crust been present from the beginning or did they become differentiated later?

The questions are difficult because no direct evidence of the earth's remote beginning is preserved in the rocks. The oldest sedimentary rocks we see postdate the origin by several hundred million years and show unmistakable signs of differentiation of core, mantle, and crust, as well as of an ocean and an atmosphere.

The history of the ocean and atmosphere has become clearer since we have observed firsthand the atmospheres of other planets. The atmospheres of Venus and Mars are primitive in the sense that they have apparently never been modified by the development of organisms. Until a generation ago many believed that the ocean and atmosphere condensed from a primordial vapor as soon as the earth's surface cooled below boiling temperatures. Today, however, the evidence against a primary origin is overwhelming and the ocean and atmosphere are believed to have been derived from the earth's rocky mantle and crust through time. The important questions now are how quickly and in what manner the fluid cover evolved. How the earth got its moon also remains a puzzle. Did the moon originate from primordial material as it orbited the earth as a satellite already in orbit? Was it captured early in geologic history? Could it possibly be a fragment of the primordial earth itself? These questions have yet to be answered.

ORIGIN OF THE SOLAR SYSTEM

The sun is an ordinary star, slightly smaller than average. Its surface temperature is 5500°C and it radiates dominantly in the yellow, which is normal for a star of its size. Like all stars, the sun evolved from a nebula—that is, an enormous cloud of widely dispersed interstellar gas and dust like the one shown in Fig. 2-1. The nebula condensed by itself under the gravitational attraction of its component particles and, as it became smaller, it also became hotter because the potential energy of its inward-falling particles was converted into heat energy. Eventually the most dense, central part of the shrinking cloud achieved a high enough temperature to emit infrared radiation; a short time afterward it began to glow visibly as contraction continued. The initial collapse of the nebula, from a diameter of perhaps a light year to a protostar with a diameter about as great as the orbit of Mercury, occurred quickly, possibly within a few decades. Subsequent contraction to the sun's present size happened much more slowly. Fifty million years may have elapsed before the core of the gradually contracting mass attained the million-degree temperatures necessary to initiate the hydrogen fusion from which it now derives its energy and to establish it as a stable star.

As the sun condensed and evolved, all the planets, satellites, asteroids, and comets that constitute the solar system condensed and accumulated, too. The details of the origin of the solar sysem are elusive, but much work in progress seeks to discover precisely how the planets, in particular, derived from the gaseous disc that was spinning about the contracting protosun. How the solar system may have originated is shown schematically in Fig. 2-2. The enormous

FIG. 2-1 A typical nebula: Massive clouds of gases and dust that are believed to give rise to stars. (Hale Observatory)

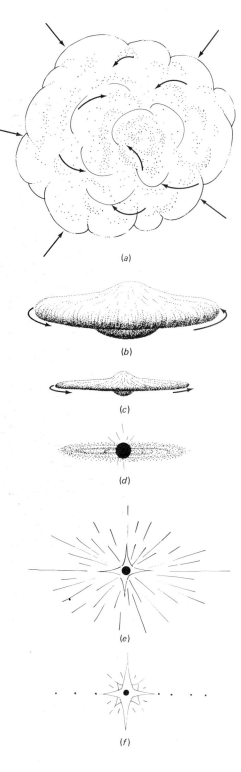

FIG. 2-2 Evolution of the solar system from a huge dispersed nebula. In the earliest stages the nebula contracts and begins to rotate. Particles condense and accumulate into the planets. As the protosun heats up, it begins to shine. Intense solar wind drives off gases from the disc surrounding the new sun, leaving only the sun and planets where there was once an immense nebula.

(a)

(b)

(c)

(d)

(e)

(f)

nebula or dust cloud from which the sun and planets evolved possessed some initial rotation. As the cloud contracted, its rate of rotation increased as would be expected according to the law of conservation of angular momentum. Simultaneously its increasing rotation caused the cloud to flatten progressively into a disc. The massive central region of the disc ultimately became the sun and the outer portion of the disc condensed into planets and satellites, asteroids and comets.

Many of the exact circumstances surrounding the origin of the solar system will never be known, but studies of the moon, planets, and meteorites in recent years have shed considerable light on what must have happened. During the early rapid collapse of the nebula the flattening disc, as well as the central protosun, must have heated up. The disc may have achieved a temperature as high as 2000°C, and because of these high temperatures, the material in the disc, although increasing in density, remained in a gaseous state. When the contraction of the nebula ultimately began to slow, the central protosun still probably continued to heat up, but the disc, where matter was far less concentrated, began to cool. As it cooled, the planets took shape. Within the cooling disc, tiny particles of matter condensed like dew, first forming liquid droplets and then tiny crystals, consisting mostly of silicate minerals and nickel-enriched iron. As the particles formed, they began to accumulate at favored distances from the central protosun where eddys concentrated them in the swirling gaseous disc. In these regions the particles literally fell together under their own mutual gravitational attraction and this process created larger particles that, in turn, grew through collisions with each other. In this way, the planets accumulated. From the very outset they were largely solid bodies.

How much time was required for the planets to form from the cooling disc? In one sense, their accumulation continues today, for tiny meteors continually add to the mass of the earth and other planets. But the present rate of accumulation is infinitesimally small. Astronomers now believe that once they began to form, the planets reached nearly their present masses quickly, perhaps within 10,000 years.

ACCUMULATION AND DIFFERENTIATION OF THE EARTH

The present-day solid earth shows a clear differentiation by density: the heaviest abundant elements — nickel and iron — are concentrated in the partially liquid core whereas the lighter abundant elements, mostly silicon and aluminum combined with oxygen, surround the core as the massive mantle and thin surficial crust. (Fig. 2-3).

There are two general theories for the origin of the core–mantle–crust separation and both are closely related to the initial accumulation of the earth from nebular material. Some workers believe that the earth accumulated as a homogeneous body and that the core formed later. In order to separate the heavier iron and nickel and concentrate them in the core, the interior of this

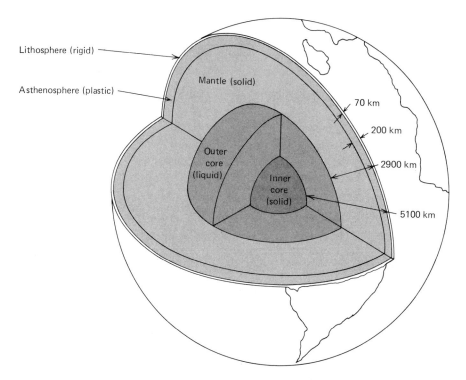

FIG. 2-3   The earth today consists of a comparatively thin outer *crust* formed chiefly of low-density silicate minerals, a thick underlying *mantle* formed chiefly of high-density silicate minerals, and a central *core* formed chiefly of iron and nickel. The upper mantle consists of a rigid outer layer that, together with the overlying crust, forms the *lithosphere*. Below it is a weak layer, the *asthenosphere*.

initially cool earth must have somehow become hot enough to melt the iron and nickel, thus allowing them to sink through the lighter, surrounding silicate minerals (Fig. 2-4). These silicates melt at higher temperatures than nickel and iron but would, nevertheless, soften and slowly flow as the earth's interior became hotter; this flow would permit the much heavier molten metals to sink and displace the lighter silicates from the earth's center.

Such a partial melting of the early earth could have been caused by heat generated through the decay of radioactive nuclides, which were present in far greater abundance early in earth history than they are today. Calculations based on estimates of the amount of radioactive material present in the initial earth suggest that soon after it formed enough heat might have been generated to begin to melt nickel and iron at depths of about 650 kilometers. These molten metals, being heavier than the soft, underlying silicate material, would tend to sink slowly as great "drops" to accumulate at the earth's center. This sinking, in turn, would create additional heat due to friction and the release of potential energy; and this additional heat would further contribute to the melting and sinking process.

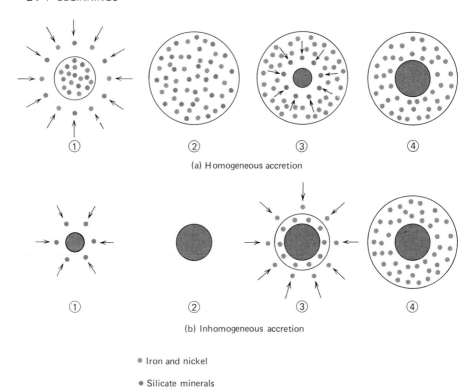

(a) Homogeneous accretion

(b) Inhomogeneous accretion

● Iron and nickel

● Silicate minerals

FIG. 2-4 Two hypotheses concerning the initial accretion and internal differentiation of the earth: (a) Homogeneous accretion. The earth accumulated as a homogeneous mixture of silicates and iron–nickel. Radioactive heating caused the iron–nickel to melt and sink to the center of the earth, where it formed the core. Remaining silicates became the mantle and crust. (b) Inhomogeneous accretion. Iron–nickel was the first phase to condense from the nebula and to accumulate to form the core. Slightly later, the silicates of the mantle and crust condensed and were accreted around the core.

As the molten metals gradually sank to form the core, the lighter silicates would have risen into the mantle. At the same time, the lightest and most easily melted elements, such as sodium, calcium, potassium, and aluminum, would have tended to move upward through the mantle silicates. As a result, they would be concentrated near the earth's surface, where they are found today in the rocks of the crust and upper mantle.

This hypothesis of the earth's formation requires a series of complex condensation processes *before* accretion began in order to produce the homogeneous mixture of metallic and silicate particles that formed the earth. Another major difficulty with the homogeneous accretion model is that the abundances of nickel, cobalt, copper, and gold in the mantle are at least ten times greater than they would be if the silicates of the mantle and metallic iron of the core had ever been in chemical equilibrium. Yet equilibrium would certainly have been achieved if they were uniformly distributed within the homogeneous earth and then slowly heated to the melting point of iron.

Other workers believe that the earth's core probably existed from the beginning. They point out that iron and nickel would be the first abundant materials expected to condense from the primordial nebula. If accumulation of the earth began as soon as solid materials condensed, then iron and nickel were potentially the first abundant materials to constitute the initial, interior part of the earth (Fig. 2-4). With further cooling of the nebula, silicates would be expected to condense in abundance and to accumulate around the iron and nickel-rich portion, forming the earth's mantle. The sharp boundary that presently separates the earth's metallic core from its overlying silicate mantle could have formed at this early time, particularly if temporary melting occurred, because metallic liquids and silicate liquids are immiscible.

This theory also has its difficulties. Calcium and aluminum oxides should condense from the solar nebula before iron and nickel. Even if the iron and nickel sank through the oxides as they accreted, the oxides should form an essentially iron-free, calcium and aluminum-rich layer at the base of the mantle. Such a layer would easily be detected on the basis of its distinctive seismic properties, but none has been found. There is also good experimental evidence that the condensation temperatures of iron and of mantle silicates overlap considerably so that initial differentiation of an iron core and silicate mantle would not be clear-cut. The model implies that the volatile-rich materials from which the atmosphere and ocean were derived accreted very late, when accretion velocities should have been high, due to the relatively large size the earth had already achieved. At such velocities volatile-rich materials would be mostly vaporized in the atmosphere or on impact rather than accreted. Finally, the core is not exclusively iron and nickel. Some 10 to 20 percent appears to be a lighter element, probably sulfur, silicon, or carbon; this theory does not offer any mechanism by which these elements could have been incorporated into the core.

In any case, the earth probably formed as a largely solid body. Had all of it been liquid at one time, then almost all the earth's iron should have sunk to the core and the crust and upper mantle would not have remained as rich in iron as they are. The true story of the earth's accretion and differentiation probably incorporates features from both models. It seems likely that the earth began to accrete from the earliest materials that condensed from the nebula, as in heterogeneous accretion. It also seems likely that the accretion of volatile-rich materials, which have subsequently been degassed to produce the atmosphere and the ocean, were accreted at an early stage when impact velocities and temperatures were low. Some researchers have suggested that under relatively cool (less than 700°C) accretion temperatures the first iron to condense was probably in an oxidized rather than a metallic phase. Later, as the mass of the earth increased, temperatures would also have increased due to higher impact velocities. At higher temperatures metallic iron (rather than iron oxide) would be more stable and would contribute the major portion of accreting materials. The stage would be set for a differentiation event in which the metallic iron and nickel would sink to the center of the still-accreting earth, thereby forming the present metallic core.

Asteroids, like the planets, also originated during condensation of the primordial nebula. A few have diameters of several tens of miles and are clearly visible through telescopes. Most occur in an orbital belt that lies between Mars and Jupiter, but some have been deflected from this belt and follow orbits that intersect those of the earth and other planets. The ones that fall to the surface of a planet are called *meteorites*. Meteorites were once thought to have come from the breakup of a planet that formerly existed between Mars and Jupiter. Today most astronomers believe that meteorites never belonged to bodies more than a few miles in diameter. Although forming at the same time as the planets, meteorites themselves appear never to have collected together into a single planet.

Meteorites provide good evidence that the initial condensate from the nebular disc was not homogeneous because the hundreds that have been studied show a wide array of compositions. They have been categorized into two groups: irons and stones. These groups are now believed to represent an extremely early crystallization of two distinct phases. Iron meteorites consist chiefly of metallic iron with 4 to 20 percent alloyed nickel and small amounts of other elements. Irons constitute only about 7 percent of meteorites that fall to earth. Stony meteorites consist largely of silicate minerals. Inasmuch as these earliest fragments of the primordial solar system are markedly differentiated, it is likely that the earth similarly formed as a partially differentiated body. Further differentiation is not precluded, however. Indeed, many geologists believe that a major heating event associated with formation of the metallic core did take place early in the history of the earth.

ORIGIN OF THE OCEAN AND ATMOSPHERE

While the earth accumulated, the region of the nebula surrounding it must have become increasingly enriched in the gaseous compounds of hydrogen, which is by far the most abundant element in the solar system, and also in the inert gases, particularly neon and helium. The inert gases could not have been incorporated in rock-forming minerals and we might expect that, together with the excess of hydrogen compounds that surely existed, they would have been entrapped by gravity to form a primitive atmosphere. If these gases were trapped, however, they subsequently escaped into space because the present atmosphere is drastically depleted in all of them (Fig. 2-5). Consequently, two questions are raised. Where did these gases go? And if the present atmosphere was not inherited from the gaseous stuff left over when the earth formed, where did it come from?

The answer to the first question is that the inert gases and excess hydrogen compounds were probably driven off into space by an intense solar wind consisting of a concentration of high velocity particles emitted from the sun.

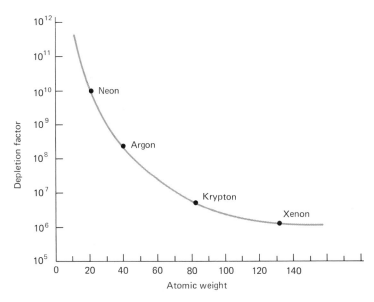

FIG. 2-5 Inert gases in the earth's atomsphere are depleted between 10^{10} and 10^6 times relative to their cosmic abundances, depending on their atomic weight. This is excellent evidence that the atmosphere is not inherited from the primordial nebula but was formed later.

The earth, along with Mars, Venus, and Mercury, which are similarly depleted in these materials, was probably subjected to a barrage of solar emissions at this early time when the sun was not yet stabilized. These intense solar winds could easily have driven away gaseous material from the inner planets just as the present mild solar wind drives the material in a comet's tail away from its head. The way in which the noble gases are fractionated (that is, the lighter the gas, the greater the loss) strongly supports this idea (Fig. 2-5). Only one place is left from which the present atmosphere — and the ocean as well — could have been derived. At the time the primordial atmosphere was being driven away, the compounds and elements that constitute the present ocean and atmosphere were actually a part of the rocks that make up the solid earth. Subsequently they have been supplied to the earth's surface by means of volcanic activity through geologic time.

Volcanoes today release large quantities of gaseous compounds from the earth's interior, most of which were previously locked in rocks. Volcanic emissions over the long course of earth history are the most probable source of the earth's oceans and atmosphere (Fig. 2-6). The composition of the earth's initial atmosphere is uncertain, but it may well have been similar to the composition of gases given off by present-day volcanoes that have their source deep within the upper mantle. Water vapor and carbon dioxide (CO_2) are the principal constituents of modern volcanic gases. Hydrogen, nitrogen, ammonia (NH_3), methane (CH_4), chlorine, and many other gases occur in smaller quantities. The water vapor released early in earth history would have condensed to form

FIG. 2-6 Mount St. Helens, May 18, 1980. Early in earth history volcanoes such as this supplied the water, carbon dioxide, and other gases from the mantle that formed the primitive ocean and atmosphere. (J. G. Rosenbaum, U. S. Geological Survey)

the initial ocean as the underlying crustal rocks cooled below 100°C, leaving a primitive gaseous atmosphere composed largely of carbon dioxide.

The ocean has probably changed relatively little in composition through geologic time, for it is still 96.5 percent water. The actual volume of oceanic water that has existed throughout earth history is less certain. Early volcanism may have released large quantities of water to produce a large initial ocean. Alternatively, the volume of water may have increased gradually through geologic time.

In contrast to the relatively constant composition of the ocean, today's atmosphere of nitrogen (78 percent) and oxygen (21 percent) differs greatly from the early atmosphere, which was probably dominated by carbon dioxide. Next to steam, carbon dioxide is the most abundant gas emitted from present-day volcanoes and it makes up most of the present atmospheres of Venus and Mars, the neighboring planets that most closely resemble the earth in size and structure. The processes that acted over the long course of earth history to produce the present nitrogen–oxygen atmosphere are principally of two types.

The first are inorganic processes of change, particularly the removal of carbon dioxide and other reactive volcanic gases from the atmosphere by chemical reactions with crustal rocks and ocean water to form new solid materials. The exact nature and the rates of these chemical reactions are still speculative, but some clues are provided by ancient crustal rocks.

The second cause of atmospheric change lies in the origin and expansion of life. Both nitrogen and oxygen, the primary gases of the present atmosphere, as well as the much smaller quantity of present-day atmospheric carbon dioxide, are involved in continuous cycles in which they are first removed from and then later returned to the atmosphere by the metabolic activities of animals and plants. Because of these cycles, the history of the changing composition of the atmosphere is closely tied to, and in large measure controlled by, the history of life.

THE BEGINNINGS OF LIFE

Although abundant fossilized remains of animal life first occur rather late in earth history, the remains of primitive plants are found in Precambrian rocks of all ages, including some that are almost as old as the oldest known rocks. Because plants have been present since the earliest recorded earth history, we conclude that the origin of life on earth, like the differentiation of the solid earth, ocean, and atmosphere, must have taken place in the dim "pregeological" interval lying between the formation of the earth and the origin of the oldest surviving rocks. Even though there is no direct evidence for life's origin, the subject has attracted much speculation and experiment. All living things are composed largely of compounds of the four elements hydrogen, carbon, nitrogen, and oxygen. These elements were undoubtedly present in large quantities in the earth's initial ocean and atmosphere, just as they are today. Biologists now believe that life arose when simple compounds containing these elements were combined into more complex compounds by ultraviolet radiation and other energy sources present on the early earth. This belief has been strengthened by numerous laboratory experiments conducted over the past 20 years that attempt to simulate the early environment in which life might have arisen. None of these experiments has produced anything even approaching the complexity of the simplest organism; yet they have clearly shown that a variety of the complex chemical building blocks that make up life could have been present in the early ocean and atmosphere. Such building blocks form readily when any of the possible energy sources is applied to almost any gaseous mixture containing the elements hydrogen, carbon, nitrogen, and oxygen.

There is, however, one strict limitation on the origin of these biological building blocks. Their production always requires gaseous mixtures in which any oxygen present is combined with carbon, nitrogen, or hydrogen rather than occurring as the free oxygen gas that constitutes about one-fifth of the present atmosphere. The chemical building blocks do not form in the presence of free oxygen. This limitation is not surprising because the abundant free oxygen of the present atmosphere has apparently been produced by the metabolic activities of green plants through the long course of earth history. Before oxygen-producing plants arose, there is every reason to believe that the earth's atmosphere, like present-day volcanic gases, lacked free oxygen.

THE MOON

Most of the nine planets have one or more satellites in orbit around them: the earth–moon system is unique only because the moon is comparatively large. The diameter of the moon is about one-quarter as large as the earth's whereas most planetary satellites have diameters less than one-twentieth the size of their respective planets.

The moon's diameter, total mass, and distance from the earth have long been established by astronomical calculations. Knowing the diameter and mass, the average density of lunar materials was long ago calculated to be 3.3 grams per cubic centimeter, about the same as the materials of the earth's upper mantle but far lighter than the earth's overall density of 5.5 grams per cubic centimeter. This situation suggests that the moon lacks the heavy iron core of the earth's interior but is, instead, dominated by silicate minerals throughout. Because it consists of lighter materials, the total mass of the moon is only about one-eightieth that of the earth.

The Lunar Surface

From afar the most prominent features of the lunar earth side are large dark patches surrounded by lighter areas; both have been observed directly by everyone, for they are readily visible to the unaided eye (Fig. 2-7). Early telescopic observations showed the dark areas to be smooth, flat regions that contrast sharply with the irregular and rugged terrain of the lighter regions around them. Because they bear a superficial resemblance to the earth's oceans, the dark, smooth areas were long ago named *maria* (Latin for "seas"; the singular is mare); the lighter regions of high relief are known as lunar *highlands*. Maria cover about one-third of the moon's earth side hemisphere, but for reasons that are still uncertain, they are almost absent from the far side (Fig. 2-7). The far side is covered almost entirely by rugged highlands.

On close inspection the margins of many of the maria show traces of the rugged relief of the neighboring highlands protruding through the flat maria surfaces and giving the impression that the mare material rests on top of an older highland surface. For this reason, geologists postulated long ago that the dark maria represent huge lava flows that spread to fill large, rugged depressions on the highland surface. This idea was confirmed by samples from the maria surfaces, which show their dominant rocks to be dark, basaltic lavas. Except at the maria margins, the lavas completely cover the older highland surface that underlies them. Judging from the exposed highlands, this buried surface probably had mountains at least 6000 meters (20,000 feet) high. The complete burial of such rugged topography indicates that an enormous volume and

 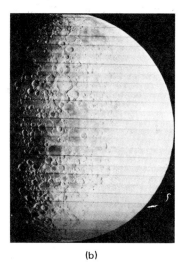

(a)   (b)

FIG. 2-7 (a) The lunar earth side showing maria (dark areas) and highlands (light areas). The dots indicate sites of samples returned to earth by Apollo 11, 12, 14, 15, 16, 17 (and Luna 16, 20) missions. (b) A portion of the lunar farside. This is dominated by highlands and lacks the large maria of the nearside. (a: Lick Observatory; b: NASA)

thickness of lava were generated to form the maria at some stage in the moon's history.

Lunar History

Long before the first lunar samples were returned to the earth in 1969, a large body of information about the moon's history had been compiled by dating the various surficial features relative to each other. Telescopic study had shown that the maria lavas that overlie the more rugged topography of the surrounding highlands are composed not of single huge lava flows but of several smaller flows that apparently formed at different times. Because the maria lack the abundant larger craters of the highlands, clearly many more large meteorites collided with the lunar surface before the maria formed than after (Fig. 2-8).

It is also possible to determine the relative ages of many large lunar craters, both those of the highlands and the less abundant ones that occur on the maria. Often one crater can be seen to be superimposed over a portion of an older one. More commonly, it is necessary to infer the relative ages of craters by their degree of modifications by later meteorite impacts [see Fig. 2-8(A)]. Using a careful combination of all such evidence of relative age, lunar scientists have been able to compile, from photographs alone, detailed geologic maps of the lunar surface showing the relative ages not only of such major features as the maria and highlands but of many individual craters and smaller lava flows as well (Fig. 2-9).

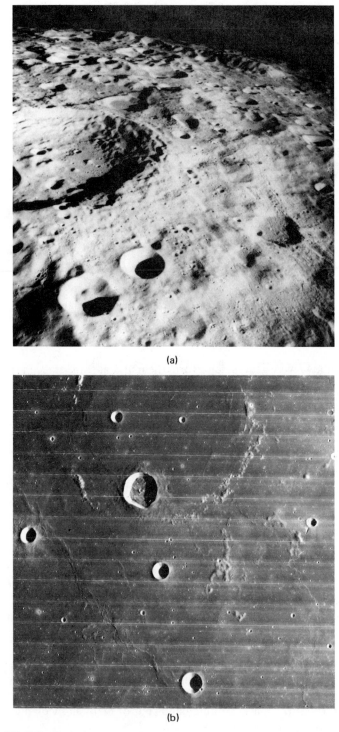

(a)

(b)

FIG. 2-8 Typical craters of the lunar highlands (a) and maria (b). (NASA)

From such studies a basic fourfold subdivision of major events in lunar history has emerged. First came the solidification and differentiation of the highland rocks, followed by massive, large-scale cratering of the highlands. During the final stages of this episode a few huge multi-ringed basins were formed by especially large impacting bodies. Next, some of the largest highland craters, including most of the multi-ringed basins, filled with the mare basalts. Subsequently, smaller-scale cratering has occurred on both the mare and highland surfaces (Figs. 2-9, 2-10).

The major events of lunar history are clearly recorded in its surface features. Yet until actual moon rocks were available for radiometric dating, no

FIG. 2-9 Lunar surface photograph with numbers showing Apollo mission landing sites (a), and a geologic map (b) compiled from it. The region shown, a part of a mare margin and adjacent highlands, contains materials formed during the four principal events of lunar history: highlands (oldest); premare craters; mare rocks; and young craters. (a: NASA; b: Mutch, 1970)

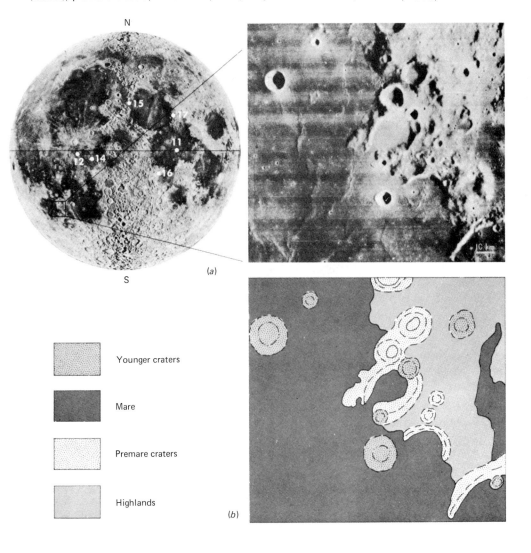

N

S

(a)

Younger craters

Mare

Premare craters

Highlands

(b)

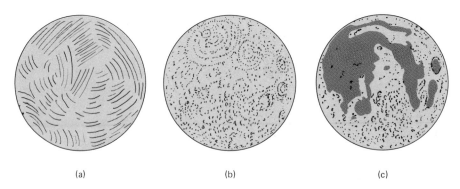

(a) (b) (c)

FIG. 2-10 Schematic diagram of the principal events of lunar history: (a) early crystaliza-tion of the anorthositic highland rock (b) cratering of highlands by large meteorites; (c) flooding of large areas by mare basalts. Only comparatively light meteorite cratering has oc-curred subsequently.

reliable means of determining the absolute ages of these events existed. Earlier estimates of the age of the maria surfaces had been made from the density of their small craters. By calculating the average number of meteorites reaching the earth today and making corrections for the moon's smaller mass and other factors, it was possible to arrive at the time required to produce a given density of craters, assuming that the rate of meteorite infall remained constant. Such estimates first suggested that the maria surfaces were very old in terms of earth history, having formed perhaps 3000 millions years ago. This conclusion has been amply confirmed by radiometric dating (Fig. 2-11).

All mare basalts so far dated have ages in the range of 3000 to 4000 million years. Thus even these relatively late features in lunar chronology formed near the time of origin of the oldest known earth rocks. Dates from the lunar highlands are only slightly older — around 4100 million years — indicating that most rocks now exposed on the moon's surface originated in a relatively short interval early in its history. Except for continuous cratering, they have remained relatively unchanged for 3000 million years or since early Precam-brian time on the scale of earth history (Fig. 2-11).

Origin of the Moon

Exactly how the moon formed constitutes a separate problem that has not yet been solved. The moon may have formed by fission of the earth; it may have formed elsewhere in the solar system and subsequently been captured by the earth; or it may have formed in orbit very near the earth.

Because the moon has no iron core and is made largely of silicate rocks, like the mantle and crust of the earth, some scientists have suggested that the moon may have originated simply from a severed portion of the earth's man-tle. In this view, the moon was originally a part of the earth, but the earth–moon body was spinning so rapidly that it quickly became unstable and broke apart [see Fig. 2-12(a)]. One problem with this earth-fission theory of

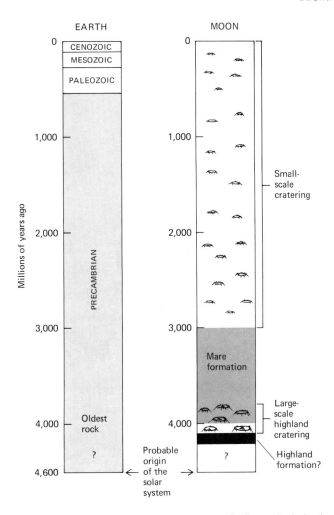

FIG. 2-11 Comparative histories of the earth and moon. Radiometric dating has shown that the principal events of lunar history took place during the early Precambrian interval of earth history.

lunar origin is that the angular momentum that the earth and moon now possess would be inadequate to trigger their separation if they were merged into the same body. Another objection is that there are major chemical differences between the earth and the moon. Samples collected by the Apollo missions indicate that, relative to the earth, the moon's surface is significantly depleted in such elements as potassium and rubidium, whose compounds volatilize at comparatively low temperatures, and it is correspondingly enriched in refractory elements like titanium and chromium. Both the angular momentum and the compositional objections might be resolved if about half the moon's mass had somehow vaporized and been lost to space by intense heating during the break from the earth. In this case, however, a particularly complex se-

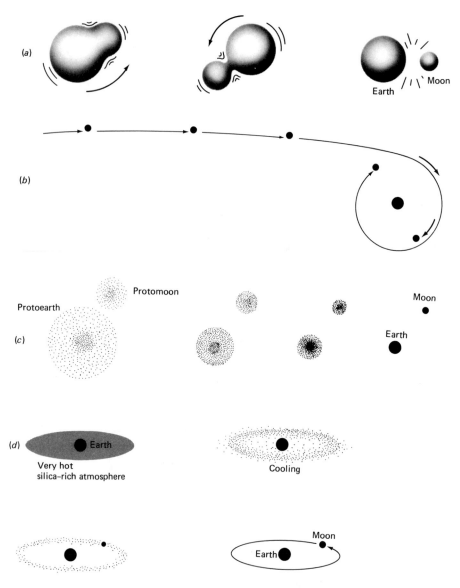

FIG. 2-12 Possible ways in which the moon formed: (a) Fission of combined earth–moon body. (b) Capture of body that was formed earlier elsewhere in the primordial nebula. (c) Independent condensation from a source near the earth. (d) Rapid formation of the moon in a distended, silicate-rich atmospheric disc around a large, hot protoearth.

quence of events would be required and so it does not seem too likely that the moon is simply a severed piece of earth material.

In order to explain the considerably different composition of the moon, some scientists have suggested that it formed in a region of the condensing primordial nebula far removed from the earth and then was captured at a much later date [see Fig. 2-12(b)]. Yet the moon's density of 3.3 is substantially below the density range (4.0 to 5.5) of the other inner planets and this density difference must reflect a considerably different composition. If the moon formed separately somewhere in the inner reaches of the nebula, it is still difficult to see why its properties are not more like those of the other planets that formed there. In addition to these difficulties, most scientists feel that capture of the moon is a statistically improbable event.

Possibly the moon formed in orbit close to the earth. Here it could condense independently from a nearby region of the nebular disc just as the earth did. If the moon condensed totally on its own, separate from the influence of the earth [see Fig. 2-12(c)], however, why does it differ so in composition, being depleted in volatile elements and lacking an iron core? A possible answer is that the moon formed in orbit but not from a separate center of accumulation at all. Perhaps instead it condensed out of a very hot, distended disc of vaporized silicates that surrounded the accumulating protoearth [see Fig. 2-12(d)]. In this view, a silicate-rich "atmosphere" could have vaporized from the earth's surface when it achieved extreme temperatures due to high-energy impact of infalling particles during the late stages of accumulation.

At the 2000°C temperatures required for vaporization of silicates, the volatile materials that are comparatively rare on the moon were particularly vulnerable to being swept deep into space by the intense solar wind that probably existed at that time. Then when the silicate vapors cooled, the less volatile materials that remained behind condensed into small planetesimals, which formed a disc analogous to the rings of Saturn but considerably more massive. The moon then rapidly coagulated from the volatile-depleted planetesimal disc, perhaps in a few hundred years.

three

the precambrian earth

The 4000 million years that elapsed between the formation of the earth and the appearance of abundant animal fossils at the beginning of the Cambrian, about 570 million years ago, are known as the *Precambrian* interval of earth history (Fig. 3-1). Rocks that formed during Precambrian time make up the bulk of present-day continents. Even though much of the continents' surface area is covered by a veneer of Phanerozoic sedimentary rocks, most of the volume of the continental crust consists of igneous and metamorphic complexes that lie beneath this sedimentary cover. These complexes of "basement rocks" are predominantly of Precambrian age. If all the Precambrian rocks were covered by younger sediments, we would know little of Precambrian earth history, for the rocks could be studied only in occasional mines or drill holes that penetrated the overlying sediments. Fortunately, each of the present continents has a relatively large area where Precambrian rocks are exposed at the surface without an overlying sedimentary cover. These are the *shield* areas (Fig. 3-2) of the continents. Most of our knowledge of the earth's Precambrian history comes from studies of rocks exposed in these regions.

THE PRIMORDIAL CRUST

The earth accreted as a relatively cool body, at least in the earliest stages. Internal heat on the primordial earth was supplied by the continued impact of accreting particles and by radioactive decay. Most geologists believe that a profound heating event occurred during the early history of the earth and this heating event is usually ascribed to the formation or final differentiation of the earth's iron–nickel core. The process of core formation was highly *exothermic*

FIG. 3-1 The Precambrian interval of earth history. The bulk of the earth's past is represented by Precambrian rocks, which span the time interval from about 4000 to 570 million years ago.

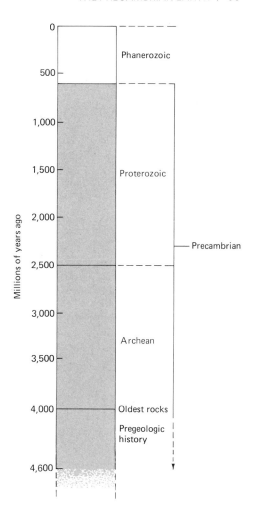

(that is, a large quantity of heat was liberated). Estimates of the quantity of heat released vary, but it seems likely that enough heat was liberated to melt substantial portions of the outer several hundred kilometers of the earth, possibly even including portions of the surface. This hypothesized heating event may have produced the primordial crust of the earth.

Various compositions have been suggested for this earliest crust, including felsic, mafic, ultramafic, and anorthositic, as on the moon. It seems most probable that large-scale melting in the mantle produced a crust of mafic-to-ultramafic composition and studies of rare-earth elements in Precambrian felsic rocks suggest that they were derived by partial melting of still older crustal materials of mafic composition. Possibly this first crust was rather unstable for several reasons. Rapid mantle convection due to high internal temperatures enhanced the recycling of crustal materials and heavy bombardment by mete-

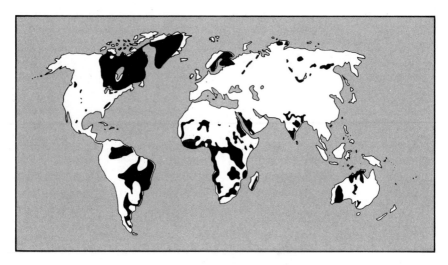

FIG. 3-2 Precambrian shield areas of the world.

orites, until about 3.9 billion years ago, caused large-scale disruption of the crust. If any of this earliest crust remains, it probably exists only as wispy fragments in the oldest high-grade metamorphic rocks or as xenoliths in Archean intrusives.

Meteorite bombardment may have had relatively long-term major effects on the early earth even though the craters and fractures that the meteorites produced have long since been erased by the inevitable forces of erosion and tectonics. During the earliest stages of bombardment the impact of meteorites must have impeded the segregation of a primitive crust by remixing the mantle and incipient crustal materials. As cooling proceeded and the crust stabilized, the major effects of bombardment were to enhance cooling and volcanic activity by making deep excavations in the crust and by the accompanying upwarping of the isotherms beneath the excavations. Evidence from the moon suggests that in the final 200 million years of major bombardment (about 4.1 to 3.9 billion years ago) the earth was impacted by perhaps 25 or 30 large bodies, 50 kilometers or more in diameter, that were capable of creating impact structures of a few hundred to a few thousand kilometers in diameter!

To understand how such impacts might have triggered the formation of protocontinents, let us look at what must have happened when one of these large meteorites hit the earth (Fig. 3-3). A 1000-kilometer impact structure could result from the impact of a meteorite 50 to 60 kilometers in diameter; the meteorite would penetrate into the earth and instantaneously explode, excavating a crater around 100 kilometers deep. The rocks surrounding the crater would be greatly heated. The instantaneous release of pressure from the removal of 100 kilometers of overburden would cause partial melting of the rocks beneath the crater, resulting in volcanic activity, which would be enhanced by the heat produced by the impact and by fracturing of the rocks.

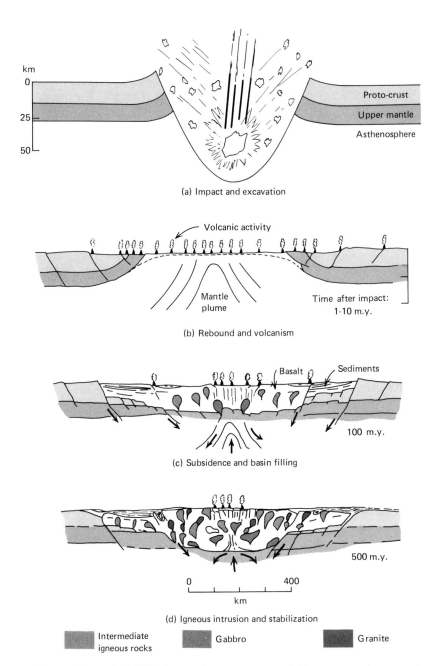

km
0
25
50

Proto-crust
Upper mantle
Asthenosphere

(a) Impact and excavation

Volcanic activity

Mantle plume

Time after impact:
1-10 m.y.

(b) Rebound and volcanism

Basalt   Sediments

100 m.y.

(c) Subsidence and basin filling

500 m.y.

0   400
km

(d) Igneous intrusion and stabilization

Intermediate igneous rocks   Gabbro   Granite

FIG. 3-3 Evolution of a 1000-kilometer impact structure. (a) Large meteorite excavates a crater about 100 kilometers deep. (b) Region rebounds isostatically, bringing high-temperature isotherms close to the surface as a mantle plume. Impact-induced volcanism breaks out throughout the region. (c) Region subsides with loading due to deposition of volcaniclastic sediments and basaltic lava flows. Melting at depth of crater-filling material produces intermediate and granitic magmas that intrude overlying basin-fill rocks. Basaltic volcanism continues and some gabbro is intruded at depth. (d) The final form is a relatively stable protocontinent consisting of a substantial portion of intermediate and granitic rocks. (The vertical scale is exaggerated four times.) (Grieve, 1980)

Within a million years, the floor of the impact structure would be uplifted by isostatic rebound. Upward distortion of isotherms by the uplift would, in effect, create a protrusion of the mantle, commonly called a "mantle plume," beneath the impact structure. The impact basin would begin to fill with lava flows and *volcaniclastic rocks*—breccias, sandstones, and tuffs consisting of volcanic debris derived locally within the basin. Eventually, the weight of infilling material would lead to gradual subsidence of the basin. The subsiding volcanics would be metamorphosed and, even though they were probably mafic in composition (Fig. 3-4), they could have been melted *partially* to form intermediate to felsic magmas that would intrude the overlying mafic basin fill. Therefore these very large impact structures may have contributed significantly to the formation of a large mass of sial (that is, crustal material composed of low-density felsic rocks rich in silicon and aluminum that characterize today's continental crust). These masses could then act as nuclei around which the earliest continents could grow. During the period from 4.6 to 3.9 billion years ago at least 25 impacts probably formed structures greater than 1000 kilometers in diameter and as much as 30 percent of the earth's surface was affected by impact structures greater than 100 kilometers in diameter. Thus early meteorite bombardment was possibly the dominant agent in the initial shaping and early evolution of the continents.

FIG. 3-4 Classification of the common igneous rocks by mineral composition. The graph shows the proportions of common minerals occurring in each rock type. The names in parentheses are fine-grained equivalents. Dark-colored iron- and magnesium-bearing minerals are shaded.

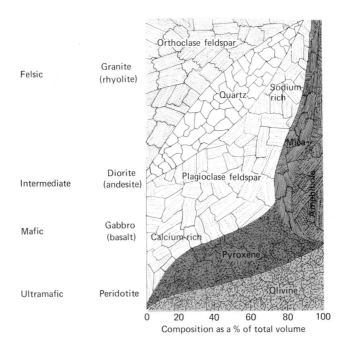

THE OLDEST ROCKS

Most exposed Precambrian terranes consist of metamorphic rocks and granitic intrusives. These rocks formed from the recrystallization or the actual melting of still older rocks within the earth's crust. Many probably represent the roots of ancient mountains. However, any mountains that may have existed in that remote time have long since been eroded away and the resulting sediments have been recycled into younger rocks, many of which have themselves been subsequently destroyed, either by uplift and erosion or by deep burial and melting. Precambrian terranes also contain significant areas of volcanic and sedimentary rocks that are metamorphosed very little. The sedimentary rocks are of special interest because they record ancient subaerial and submarine environments and the earliest evidence of life. Many sedimentary sequences are several thousands of meters thick.

Because Precambrian sedimentary sequences contain no useful fossils, it has been extremely difficult to fit them into a time framework. Generally we recognize an early Precambrian time called the *Archean* and a late Precambrian time called the *Proterozoic*. But beyond that there is no agreement as yet on a worldwide Precambrian time classification. The scanty knowledge we do have of Precambrian chronology has come largely from radiometric age determinations on igneous and metamorphic rocks. These determinations reveal a long and eventful record of igneous activity, metamorphism, and orogeny (mountain-building activity).

In every Precambrian shield region of the world the ages of the rocks fall into large geographic groupings, each with distinctive structural characteristics and fold directions. These "structural provinces" are separated by sharp metamorphic or fault boundaries. Within each province, the igneous and metamorphic rock ages cluster around the date of the most recent orogenic episode. These events serve as the framework for regional classifications of the Precambrian.

Radiometric dating has also contributed enormously to our understanding of the thick sequences of sedimentary rocks of Precambrian age. Although more difficult to date than igneous and metamorphic rocks, their ages are ascertained from interbedded volcanics, from intrusives that cut them, and from structural relationships (that is, whether they formed before or after an orogeny whose age is known).

Sedimentary rocks of Archean age—older than about 2500 million years—are monotonously similar wherever found. They reflect considerable orogenic activity and appear to have formed mostly in marine environments at a time before continents had developed the stable interior portions, or *cratons*, that have remained little changed throughout later geologic history. Most of these ancient rocks are metamorphosed and strongly folded and faulted. Consequently, they are difficult to study. Nevertheless, these Archean rocks are of great interest because they reflect formative stages of the earth's crust prior to the establishment of stable continents and so they differ fundamentally from later records of geologic history.

Greenstone Belts

The best-preserved and least-metamorphosed Archean sedimentary rocks occur in thick sequences that include mildly metamorphosed volcanic rocks called *greenstones*. Archean greenstones and their associated sedimentary rocks occur in all Precambrian shield areas in huge elongated downwarps that are referred to as *greenstone belts*. Greenstone belts lie between, and wrap around, much more extensive domelike batholiths of massive Archean granite (Fig. 3-5). On the Canadian Shield, greenstone belts are concentrated in large tracts several hundred kilometers in length and several tens of kilometers across (Fig. 3-6). Between the broad greenstone belt tracts lie equally vast tracts of granite and high-grade metamorphic rocks.

The most common Archean sedimentary rocks are greywackes (poorly sorted sandstones) that are typically interbedded with shales. Individual beds are thin, rarely more than 1 meter in thickness, and commonly graded—that is, grain size in each bed becomes finer from bottom to top (Fig. 3-7). Grading, which is common in greywackes of all ages, results from the rapid deposition of a bed by a single, rapidly moving, sediment-laden current (called a "turbid-

FIG. 3-5 The Barberton (Swaziland, Africa) greenstone belt showing included and adjacent granite plutons. (Windley, 1976)

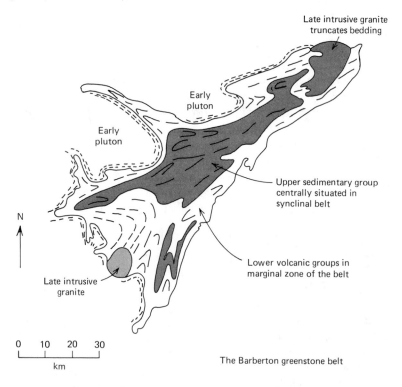

Late intrusive granite truncates bedding

Early pluton

Early pluton

Upper sedimentary group centrally situated in synclinal belt

N

Lower volcanic groups in marginal zone of the belt

Late intrusive granite

0   10   20   30
km

The Barberton greenstone belt

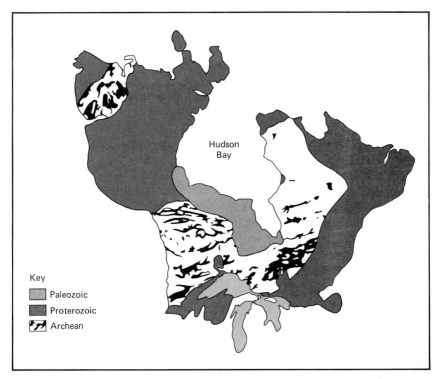

FIG. 3-6 Archean greenstone belts (in black) of the Canadian Shield. (Baragar and McGlynn, 1976)

ity current") that moves downslope under the influence of gravity and that deposits its load, coarsest material first, as it slows and finally stops far out on the basin floor (Fig. 3-8). Beds deposited by this mechanism are called *turbidites*. Turbidites occur throughout the geologic record, but in the Archean they dominate all sedimentary sequences. Most turbidites represent relatively deep-water deposition. Today they are being deposited blanketlike in the deep ocean basins where they help to create the smooth topography of the abyssal plains. Deep basins bordering rapidly rising source areas are especially conducive to turbidite formation. Turbidity currents are commonly triggered by submarine slumps high on the basin margin, which, in turn, may be caused by storms or earthquakes.

In addition to sandstone and shale beds, units of coarse conglomerate, commonly with thicknesses up to several hundred meters, occur in Archean greenstone belts. Many such conglomerate units persist for several kilometers on outcrop and probably represent debris flow deposits from nearby volcanic source areas. In some areas the massive debris flow deposits grade laterally into turbidites, which represent somewhat lower energy deposition farther away from the source. The only other sedimentary rock type that is common in

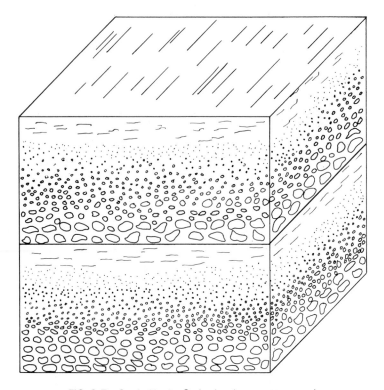

FIG. 3-7 Graded beds. Grain size decreases upward.

the Archean is a unique kind of iron-rich chert called "banded iron formation." Banded iron formations were deposited during periods of relative volcanic and tectonic quiescence, in both shallow and comparatively deep marine environments. Various other depositional environments are represented in greenstone belts, including shallow marine sandstones and deltaic and alluvial clastics, but they are relatively rare. The volcanic rocks in most greenstone belt sequences include abundant *pillow lavas*, which indicate subaqueous extrusion, and large quantities of volcaniclastic rocks.

Careful study of greenstone belts in many areas of the world has shown them to be similar. First, all greenstone belt sequences are thick, typically ranging from 6000 to 18,000 meters. Secondly, their sedimentary rocks are similar, everywhere consisting dominantly of turbidites. Thirdly, sedimentary rocks do not occur in the lowermost portions of most greenstone belt sequences, but they become increasingly common upward and may dominate in the upper portion. Finally, the various kinds of igneous rocks found in greenstone belts tend to occur in the same general succession with *ultramafic* and *mafic* flows or sills in the basal portion, andesite and other *intermediate* igneous flows and volcaniclastics in the middle, and *felsic* volcaniclastics in the upper portion.

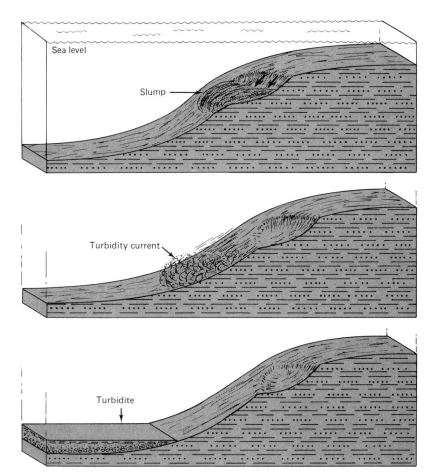

Sea level

Slump

Turbidity current

Turbidite

FIG. 3-8 Graded bedding, which is common in Archean greywackes, results from cyclical changes in transport energy. The coarsest material is deposited first, from a high-energy current. As the current wanes, finer sediments can settle out.

Young (less than 3 billion years) greenstone belts differ from old ones in that they commonly repeat the ultramafic and mafic to felsic sequence several times. The ultramafic and mafic, intermediate, and felsic succession strongly suggests progressive differentiation of a magma source at depth.

The composition of the primitive crust on which the greenstone belts were deposited is unknown. Some geologists believe that the ultramafic and mafic rocks at the base of greenstone belts represent surviving fragments of a thin, formerly widespread crustal layer that had the composition of *sima* (that is, crustal material composed of moderately dense mafic rocks rich in silicon and magnesium that characterize today's oceanic crust). Others believe that before the oldest greenstone belts formed, the earth already had a thin crust of sial and that the greenstone belts were produced on it by localized volcanism and orogeny. Indeed, the oldest rock known is a 4-billion-year-old granitic gneiss, which indicates that at least local differentiation of sialic material must have occurred very early. Actually, all greenstone belts may not have been deposited on crust of the same composition. Greenstone belts older than about 3 billion years differ significantly from those that are younger in that ultramafic and mafic volcanic rocks form a major portion of the rock sequence and intermediate and felsic rocks are rare. Volcaniclastics constitute most of the sedimentary rocks, although extensive chert beds precipitated during lulls in volcanism. Apparently no major uplifts of adjacent crustal rocks occurred in association with the development of the basins. Detrital grains of plutonic and metamorphic origin are lacking, which suggests that the crust on which these greenstone belt sequences were deposited was simatic.

In contrast to the older greenstone belts, volcanic rocks in most greenstone belts younger than about 3 billion years are separated by intervals of shales, sandstones, and even conglomerates, which are clearly derived from sialic sources. Ultramafic rocks are less abundant and intermediate volcanic rocks are more abundant than in the older greenstone belts. Thus the younger greenstone belts appear to have formed in an orogenic setting adjacent to or possibly on sialic crust.

The tectonic setting in which greenstone belts developed has been the object of much speculation. Today most tectonically active volcanic regions occur on plate boundaries, both spreading centers and subduction zones, and many geologists have speculated that Precambrian greenstone belts formed in similar settings. Others have questioned the applicability of plate tectonic processes that are operative today to the early Precambrian. In order to evaluate these ideas further, it is appropriate to digress briefly here and outline the basic principles of plate tectonics. Then we shall return to the Precambrian story and give one example of how the plate tectonic model might help to explain the setting of the greenstone belts.

Period of time 600 million — onwards years ago.

Plate Tectonics: Applicability to the Precambrian

The earth's surface today appears to consist of seven huge lithospheric plates, each with dimensions of several thousand kilometers (Fig. 3-9), and about 20 smaller plates. All move independently as coherent units. The lithosphere, of which these plates are made, includes the rigid outer 100 kilometers or so of the earth, and this area constitutes the uppermost part of the earth's mantle as well as all the crust. The lithosphere rests on a nonrigid, partly molten zone in the mantle called the asthenosphere. No one knows what causes the lithospheric plates to move relative to one another, but possibly they are carried along piggyback style by the slow plastic flow of the asthenosphere. The plates' movement requires an enormous quantity of heat energy, which comes from within the earth.

btw lithosphere + mantle.

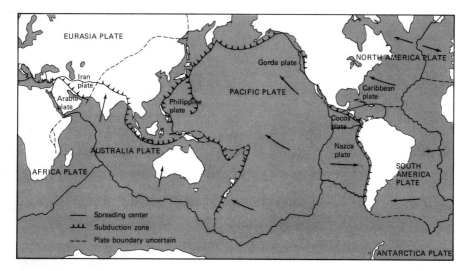

FIG. 3-9 Today 7 major and 20-odd minor lithospheric plates make up the earth's crust (only a few of the smaller plates are shown). Arrows show the inferred relative motions of the plates. White areas represent present-day continents.

Convergent   Divergent.   transform

There are three kinds of plate boundaries: those where two plates are moving apart, those where they are being pushed together, and those where plates are simply slipping sideways, one past the other (Fig. 3-10). Regions

FIG. 3-10 Schematic diagram of the three types of plate boundaries. At spreading centers new crust is added to the edges of existing plates. Subduction zones and oceanic trenches develop where two plates are coming together. Here old crust descends into the mantle and is melted. Plates slip sideways past each other along segments of transform faults.

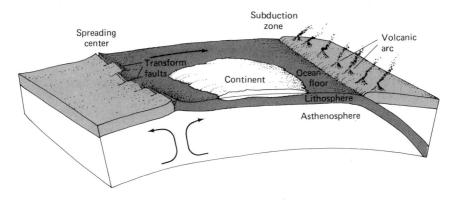

49

where plates are actually diverging, commonly called *spreading centers*, coincide with oceanic ridge–rise systems and with large-scale cratonic fissures, such as the Red Sea and the Gulf of California. As two lithospheric plates move away from each other, the zone that they vacate is filled in from below with new crust of the oceanic type, consisting of basalt flows and gabbroic and peridotitic intrusives, which probably are derived from the asthenosphere. A new ocean basin is created in the process. In regions where two plates are converging a corresponding quantity of oceanic crust is consumed as one plate is "subducted" below the other and descends deep into the mantle, where its leading edge is progressively melted. These *subduction zones* are usually sites of oceanic trenches and are typically flanked by volcanoes in the form of an island arc, such as the Aleutians, or a mainland volcanic chain, such as the Andes. The third type of plate boundary, where crustal plates are slipping past each other horizontally, are marked by faults with very large horizontal displacements. The well-known San Andreas Fault in California is an example of such a *transform fault*.

Because of its low density, sialic continental crust (as opposed to simatic oceanic crust) is rarely subducted. Indeed, this fact suggests that essentially all the sialic crust that has ever formed still exists and so we may be able to find and study the oldest sialic crust. The oldest simatic crust, on the other hand, has almost certainly been destroyed by subduction. A continent that encounters a subduction zone acts as a buoyant buttress and subduction of the plate containing the continent generally ceases. If the leading edge of the formerly overriding plate is made of oceanic crust, it now becomes the subducted plate and begins to pass beneath the continental block and to descend into the mantle (Fig. 3-11). Subduction at continental margins causes intense crustal deformation and is believed to be responsible for marginal fold mountain chains, such as the Alps and Appalachians. If, on the other hand, the formerly overriding plate's leading edge is also made of continental crust, the collision of continental masses causes much folding and thrust faulting and tends to stabilize both plates along a weld or *suture zone* between them. Collision of continental masses also causes intense crustal thickening and is believed to be responsible for interior mountain chains, such as the Urals and Himalayas.

Only oceanic crust is continually produced at spreading centers and destroyed at subduction zones. As a result, all oceanic crust is geologically young. Figure 3-12 shows the large portion of existing oceanic crust that has been produced within the past 75 million years. In fact, the oldest rocks known from the present ocean basins are less than 200 million years old. Surrounded by ocean basins that continually change shape, continental blocks are sporadically split, shoved about, sutured together, and generally battered along their margins by plate motions, but they are not destroyed.

The geologic features generated by modern plate tectonics processes have been briefly described. Did these same processes operate in the Precambrian? Can we use our modern plate tectonics model to help us understand the origin of Precambrian rocks? The answer is that we really do not know. It seems like-

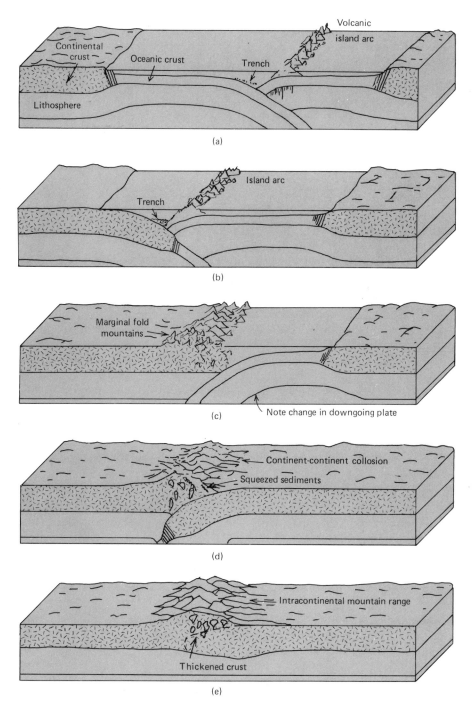

FIG. 3-11 Closure of an ocean by plate convergence. (a) Subduction of oceanic crust and development of an island arc. (b) Encounter of continental crust and trench. (c) Switching of direction of subduction and development of fold mountains at continental margin. (d) Continent–continent collision and formation of intracontinental mountain chain. (e) Cessation of convergence. Isostatic uplift.

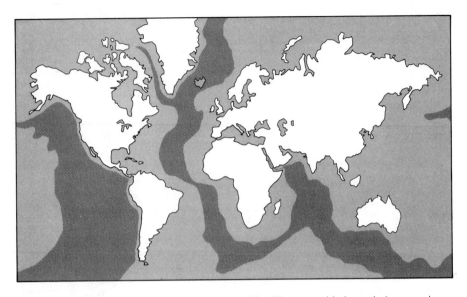

FIG. 3-12 Areas of sea floor created in the last 75 million years (darker color) cover a large part of the ocean basins. In contrast, little continental area has been formed during this time. (McKenzie, 1972)

ly that at least some of the basic concepts of the plate tectonics model can be applied to the Precambrian, but the actual modes of expression of plate tectonics processes and the preserved record left by them may be quite different from our modern examples.

In the Archean large quantities of heat were liberated by the formation of the earth's iron–nickel core, levels of radioactivity were relatively high, and massive meteorite impacts liberated tremendous amounts of heat and brought deep-seated isotherms close to the surface. This higher thermal gradient probably caused much more vigorous plastic flow in the earth's mantle than exists today. Possibly the lithosphere was much thinner (less than 50 kilometers, compared to today's 100 kilometers) and broken into relatively small plates, whose edges coincided with the boundaries between convection cells in the mantle. Until about 3.9 billion years ago these small, thin plates were frequently disrupted by the impact of very large meteorites.

Spreading centers in the Precambrian may have been fairly similar to those of today, although the spreading rates were probably greater and the volcanic activity more vigorous. Convergent plate margins, on the other hand, may have behaved quite differently than those of today. With thinner lithosphere, broad-scale buckling rather than fracturing and thrust faulting may have been a more common response to large-scale compressive stresses than it is with today's relatively thick and rigid lithosphere.

In response to increasing pressures, the descending basaltic slab in a modern subduction zone undergoes a transformation to a different kind of rock called eclogite. The density of eclogite is greater than that of basalt or the surrounding mantle and so it is this transformation that allows the descending slab to continue sinking into the mantle at a relatively steep angle. Experimental evidence, however, indicates that this transformation probably could not have taken place in the Archean because the thermal gradient in the upper mantle and crust was too high. Support for this conclusion comes from the fact that no eclogite of Precambrian age has been found. Without the transformation to eclogite, the descending slab at early convergent plate boundaries must have gone down at a relatively low angle. Consequently, arc-type magmatism would have occurred over a much broader area than it does today. It has been suggested that the vast tracts of low metamorphic grade granitic terrain between greenstone belts may have originated in this way.

Returning to greenstone belts, attempts have been made to relate them to a variety of plate tectonics settings, including spreading centers in narrow oceans, island arcs, and back-arc basins. Chemical data are equivocal: the oldest volcanic rocks of many greenstone belts have chemical compositions similar to rocks in modern spreading centers whereas the chemical aspects of the stratigraphically higher (younger) volcanics are more like those of modern island arcs. Today back-arc basins are the only setting in which the concentrations of chromium and nickel are relatively high, as they are in greenstone belts. The back-arc basin model is also appealing because it is a tectonic setting that is most likely to have been preserved. Oceanic ridges and crust tend to be subducted and island arcs are commonly reduced to remnants of their plutonic roots because the associated uplift exposes the overlying volcanic rocks to extensive erosion.

In the back-arc basin model (Fig. 3-13) the mafic and ultramafic volcanics are extruded in an opening back-arc basin where the crust has thinned. The adjacent continent and volcanic island arc supply detritus and intermediate volcanics. Eventually the basin begins to close and develops the characteristic synclinal shape as deformation proceeds. Sialic plutonic rocks derived from melting of the subducted plate beneath the arc intrude the basin complex during and/or after closing.

Some greenstone belts, especially several younger examples in the Canadian Shield, possibly formed in marginal back-arc basins. Others may have developed in different tectonic settings, some of which may have been unique to the high heat flow, and small, thin lithospheric plates of the Archean.

Archean Granitic Rocks

The granites that now surround and engulf the greenstone belts form the thickest and most stable parts of the continental crust. During the late phases of greenstone belt deposition these granites were emplaced as huge batholiths that welled upward from depth and surrounded the greenstone belts (Fig. 3-5). In this way, much of the earliest stable cratonic crust gradually accreted from below to form a largely granitic layer that exceeds 40 kilometers in thickness and that has been little disturbed since.

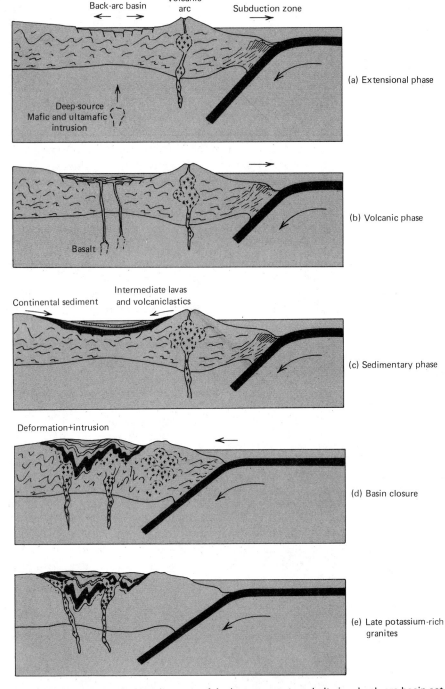

FIG. 3-13 Model for the development of Archean greenstone belts in a back-arc basin setting. (a) Back-arc extension produces a thinned and rifted crust. (b) Mafic and ultramafic volcanics are extruded through rifts in basin. (c) The adjacent volcanic arc and continent supply calc-alkaline to felsic volcanics and sediments to the basin. (d) and (e) Closure of the basin results in deformation to synclinal form and intrusion of granites. (Tarney, Dalziel, and deWit, 1976. Reproduced by permission of J. Wiley)

Archean granitic material was probably derived from the upper part of the mantle or from subducted or deeply buried mafic or ultramafic crust. The earth's mafic and ultramafic rocks contain a proportionately minor but volumetrically huge quantity of felsic minerals (chiefly quartz and feldspar). Felsic minerals not only have lower densities than mafic minerals but also melt at lower temperatures. Partial melting can thus provide a source for felsic magmas from mafic rocks. Whenever primitive mantle and/or subducted mafic crustal rocks melted, felsic minerals liquefied first. Once produced, the low-density granitic liquid could have readily ascended into the thin crust overhead to form the Archean granitic batholiths.

Greenstone belt sedimentary rocks that were intruded by the Archean granites were metamorphosed outward only for 1 kilometer or so whereas the metamorphosed zone surrounding similar Phanerozoic granitic intrusives is typically several kilometers thick. The intruding Archean granitic magmas were not cooler than younger magmas, but apparently they intruded to such extremely shallow depths that the heat dissipated rapidly. This condition may have been a result of the thin Archean lithosphere. The granitic batholiths probably formed at a much shallower depth than they do today.

MOBILE BELTS, CRATONS, AND OCEAN BASINS

The earth's continental cratons probably began early in the Archean as small sialic masses or "protocontinents" that were differentiated from mantle material below. Some geologists believe that the first protocontinents enlarged systematically by peripheral accretion of successive belts of newly formed granitic crust. Others believe that numerous small protocontinents formed piecemeal and joined together more or less randomly in time. Radiometric dating may eventually show which idea is more nearly correct, but radiometric dating of the Archean is especially difficult because intense worldwide thermal activity at the end of the eon, around 2500 million years ago, widely obliterated earlier dates. The few surviving earlier Archean dates suggest that most of the thick Archean granites were produced over a period of about one billion years, from about 3500 million to 2500 million years ago.

Platform Sequences and Geosynclinal Belts

Following the worldwide episode of igneous activity that closed the Archean, the continental cratons became stabilized. The effect on the sedimentary rock record was dramatic. The repetitious turbidites and volcaniclastics that dominate the Archean give way in the Proterozoic to a wide variety of sedimentary rock types. This variety reflects a host of sedimentary environments, some marine, some continental, and some transitional environments, such as beaches, tidal flats, and deltas. Except that they lack fossils, Proterozoic sedimentary rocks look like Phanerozoic sedimentary rocks.

Post-Archean sedimentary rock assemblages can be placed in one of two fundamental categories — *platform* and *geosynclinal*. Platform sequences form in the stable interior portions of cratons. Since the early Proterozoic the elevation of the cratons appears to have been close to sea level. At times the cratonic platforms subsided gently and were covered by extensive epicontinental shallow seas, sometimes called *epeiric* seas. At other times they emerged slightly and were shaped by subaerial processes. Unconformities are numerous in these stable platform environments and total sedimentary accumulations are generally thin — on the order of 1000 or 2000 meters. Locally, however, strata thicken in broad equidimensional cratonic *basins* where thicknesses may exceed 5000 meters. The bulk of the Proterozoic sedimentary record now preserved on the cratons was probably produced in such basins.

The margins of cratons are less stable than their interiors. During episodes of subsidence cratonic margins have commonly received sedimentary thicknesses in excess of 10,000 meters. These thick accumulations of sedimentary rocks, which commonly form in belts several hundred kilometers long, are called *geosynclines*. In any transgression of the sea these marginal areas are the first to be covered and during regression they are the last to be exposed. As a result, unconformities are fewer and the historical record is more complete in geosynclines than in platform sequences.

Geosynclinal deposition begins as a new ocean opens and sediments are shed seaward toward the spreading center to produce a thick sedimentary wedge at the trailing margin of the continent (Fig. 3-14). The rocks of a geosyncline typically represent two distinctive suites which divide the geosyncline into a *miogeosyncline*, along the inner margin, adjoining the craton, and a *eugeosyncline*, which lies seaward (Fig. 3-14). Miogeosynclinal sedimentary

FIG. 3-14 Block diagram of late Proterozoic and earliest Paleozoic miogeosyncline in Nevada and Utah, following inferred continental separation. Thickness of the strata reaches 7600 meters at the inferred shelf edge. (Stewart, 1972)

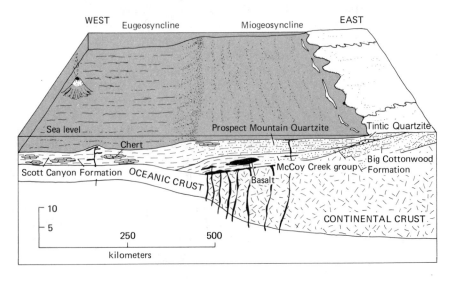

rocks are deposited on the continental shelf. These deposits represent a greatly thickened seaward extension of the same kinds of rocks that occur in the interior of the craton—shallow-water limestones, dolomites, shales, and clean, well-sorted sandstones—all of which formed in tectonically stable settings comparatively free from igneous activity. Eugeosynclinal sedimentary rocks represent contemporaneous turbidites and pelagic sediments of the continental rise and the abyssal plain. They consist mainly of shales and poorly sorted sandstones. Carbonate rocks are rare in eugeosynclines, but bedded cherts are common. Volcanic rocks are also common in eugeosynclinal suites, which probably reflects their proximity to plate margins.

Throughout geologic time trailing continental margins have commonly been changed into convergent margins. After a subduction zone develops, first the eugeosynclinal, and then the miogeosynclinal deposits are compressed in large-scale folding and thrust faulting, and then uplifted into marginal mountain chains. The roots of the mountains produced during these orogenic episodes are simultaneously intruded by granitic batholiths. These active regions at the margins of the cratons, which at times subside deeply to receive sediments of geosynclinal thicknesses and which at other times undergo large vertical and horizontal movements accompanied by igneous intrusion, are called *mobile belts.*

What would happen if the modern passive eastern margin of North America developed into a mobile belt? At present, the North Atlantic Ocean is widening as the European and North American plates move apart. A thick prism of sedimentary rocks has been accumulating at the continental margin since the Triassic, when the separation began. Suppose that today's plate motions were to reverse and the Atlantic became a closing ocean. If a subduction zone formed along the Atlantic margin of the United States (Fig. 3-15), an oceanic trench, flanked by volcanoes, would be created there. The drag of the descending lithospheric plate would slowly crumple the continental rise sedimentary prism while the volcanoes contributed their lava flows and volcaniclastics to the depositional sequence within the newly created eugeosynclinal foldbelt. With continued subduction and thrust faulting, the thick continental shelf sedimentary prism would also begin to deform into a parallel miogeosynclinal foldbelt. At depth, the selective melting of the lighter portion of the descending lithospheric plate would produce magmas that would ascend through the overriding plate and give rise to the volcanic arc. With continued subduction and deformation of the crustal rocks and with still more melting, granitic batholiths would intrude the roots of the foldbelt at depth. Further deformation would be accompanied by vertical uplift and a young mountain range would develop at the site of the present continental shelf.

A distinctive suite of rocks that has been termed "the ophiolite suite" provides evidence of ancient subduction zones. Ophiolites are thrust slices of oceanic crust that are scraped off the descending plate at the inner wall of a trench and incorporated into the crumpling eugeosynclinal belt (Fig. 3-15). The rocks of the ophiolite suite occur in an overall stratigraphic order. At the

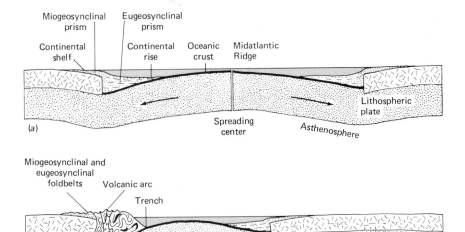

Miogeosynclinal   Eugeosynclinal
prism   prism

Continental   Continental   Oceanic   Midatlantic
shelf   rise   crust   Ridge

Lithospheric plate

(a)   Spreading   Asthenosphere
                                   center

Miogeosynclinal and
eugeosynclinal
foldbelts   Volcanic arc

Trench

Granite

(b)   Subduction
                 zone

FIG. 3-15   (a) Present-day widening Atlantic Ocean. (b) If the Atlantic began to close and the western margin became a subduction zone, the descending lithospheric plate would crumple the continental rise and continental shelf into a eugeosynclinal and miogeosynclinal foldbelt. (Dietz, 1972)

bottom are serpentinized peridotite, gabbro, and basaltic pillow lavas, which represent the igneous oceanic crust. They are overlain by radiolarian cherts and turbidites that were deposited subsequently on the spreading ocean floor. These unusual and distinctive lithologic assemblages mark the locus of many ancient convergent plate boundaries.

During the Phanerozoic, crustal shortening in response to compressive stresses and the extrusion and intrusion of large quantities of igneous rocks created high mountain ranges at convergent plate margins. Early in the Precambrian, however, high temperatures at relatively shallow depths probably prevented the formation of deep crustal roots to support high mountains; consequently, topographic relief on the surface of the early crust must have been relatively subdued except perhaps in large new impact craters. Later in the Precambrian, as radioactivity waned, mountainous terrain probably became higher and more common.

EVOLVING PRECAMBRIAN ENVIRONMENTS

The stable cratonic platforms that first appeared in earliest Proterozoic time were sites of both shallow marine and, for the first time, widespread continental depositional environments. Most Proterozoic rocks preserved today, like those of the western United States, were deposited in geosynclines or in cratonic basins. Proterozoic sedimentary rocks were at one time almost certainly more widespread but were largely eroded away late in the Precambrian when the continents became widely emergent. A large unconformity separates underlying Precambrian rocks from overlying Phanerozoic strata in most continental regions. The most common Proterozoic marine rocks that remain are sandstone and shales. In addition, the Proterozoic record is unique in containing abundant banded iron formations and a large variety of stromatolitic carbonates. Continental rocks include conglomerates and sandstones and, surprisingly, tillites (lithified glacial till) that testify to widespread Precambrian continental ice sheets.

Tillites

Only a few thousand years ago about 30 percent of the earth's land surface was covered by continental glaciers and much of this ice still remains today. If not for this fact, we would probably find it difficult to interpret thick, unstratified, poorly sorted conglomerates that are widespread in the Precambrian. Like modern tills, these ancient sedimentary rocks commonly contain boulders that are not in contact with one another but that are totally surrounded by a matrix of silt and clay (Fig. 3-16). Precambrian boulder conglomerates that are inferred to be tillites have been found on almost every continent.

Mudflows also produce unstratified and completely unsorted boulder-laden sediments that are very similar to tillites. Mudflows consist of water-

FIG. 3-16 Precambrian tillite from the Headquarters Schist of southern Wyoming.

saturated mud and rock that quickly move downslope under the influence of gravity. They are sudden, short-lived events that occur only occasionally. As a result, few geologists have observed mudflows in progress. Nevertheless, they transport huge quantities of material and in populated areas can be tremendously destructive. Today mudflows occur commonly in regions of weak volcanic rocks and in arid regions where vegetation is slight. In the Precambrian, however, before any vegetation mantled the earth's surface, they were probably much more frequent and widespread. Deposits of mudflows and tillites are nearly impossible to distinguish and many of the tillite-like rocks in the Precambrian are almost certainly mudflow deposits. But just how many is a question on which views differ widely.

Nevertheless, numerous Precambrian boulder conglomerates are universally accepted as tillites because the rock surfaces on which they rest are polished and marked with parallel scratches and grooves exactly like those in glaciated regions today. Associated with many of the tillites are well-bedded shales and siltstones that contain scattered cobbles and boulders that appear to have been dropped in from above. These strata are interpreted as marine (or in some cases lake) deposits and the "dropstones" they contain are believed to have come from melting icebergs as they passed overhead (Fig. 3-17). Similar deposits are forming today in iceberg-infested waters along the Alaskan coast and in other regions. Thus these distinctive dropstone-bearing rocks also provide excellent evidence for regional glaciation.

The widespread evidence for Precambrian glaciation defines two major episodes — the first in the early Proterozoic around 2200 to 2300 million years ago and the second in the late Proterozoic after an inferred continental break-up produced the late Precambrian margin of the western United States, around 700 million years ago. Although the dates have been ascertained only in recent years, the existence of Precambrian glaciation has been known since the middle of the last century. Tillites provide an explicit record of paleocli-

FIG. 3-17 Large and small granitic dropstones in varved shale, early Proterozoic Gowganda Formation, southern Ontario. Scale is in centimeters and inches. (Canadian Geological Survey)

mates, for they document extended periods of freezing temperatures. Even though these deposits were probably produced in high latitudes, they show that the earth was generally as cold in the early Proterozoic as it is today. Evidence for the Precambrian glaciations came as a huge surprise in the mid-nineteenth century when it was first discovered because the earth was then believed to be gradually and progressively cooling following a molten origin. Such evidence for ancient glaciation showed that, far from recording hotter temperature with increasing age, the old rocks testify that for a very long time the earth's climate was much the same as it is at present.

The Precambrian Ocean and Atmosphere

Recall for a moment that the oldest Archean sedimentary rocks consist of particles that were weathered from preexisting rocks and deposited in fairly deep water. In other words, even before the earliest rocks formed, an atmosphere and an ocean already existed. Today's ocean is 96.5 percent water and, in this repsect, the early ocean probably differed little from its modern counterpart. The large quantity of bedded cherts (massive, noncrystalline quartz) and the small quantity of carbonate strata in the early Precambrian, however, suggest that the early ocean was slightly acidic (pH less than 7) whereas today it is slightly alkaline (pH around 8). Also, the earliest ocean must have contained vastly more iron than it does today, judging from the abundance of banded iron formations in Archean and lower Proterozoic rocks. Later, in the middle and late Precambrian, when carbonate rocks become common in the sedimentary record, they are made almost entirely of dolomite, $CaMg(CO_3)_2$, instead of calcite, $CaCO_3$, which dominates in recent geologic history. This factor suggests that Precambrian oceans may have contained more magnesium than modern ones. Yet overall changes in the *total* composition of the world's oceans have probably been relatively small.

Unlike the early ocean, the early atmosphere differed greatly from its modern counterpart. In the Archean and the early Proterozoic the atmosphere contained no free oxygen (O_2). This fact is not surprising because free oxygen is chemically too active to exist in nature unless continually replenished, as it is now by photosynthesizing plants. In addition, laboratory experiments have taught us that the molecular "building blocks" of biological systems do not form in the presence of free oxygen. Thus life could never have arisen if the early atmosphere had contained free oxygen. The lack of oxygen in the early Proterozoic atmosphere is also indicated by the worldwide occurrence, in sandstones older than about 2000 million years, of detrital grains of uranium minerals (chiefly uraninite) and of fresh pyrite grains. Detrital uraninite and fresh pyrite grains are rare in modern sediments because they are so easily oxidized. Terrestrial red beds, which today are produced by oxygen-bearing groundwaters, do not occur in sequences older than 2000 million years. They do, however, occur in younger Proterozoic sequences and throughout the Phanerozoic.

Banded Iron Formations

Banded iron formations (Fig. 3-18) are extensive sequences of bedded cherts, commonly hundreds of meters thick. Today these rocks constitute the greatest iron ore reserves in the world. Banded iron formations consist of alternating red or black (iron-rich) and gray (iron poor) layers or "bands." The layers range from less than 1 millimeter to several centimeters in thickness and the iron generally constitutes about one-third of the rock. The iron usually occurs as hematite, magnetite, siderite, or pyrite and analysis of lateral relationships suggests that the oxide-carbonate-sulfide iron facies transitions represent a shallow-to-deep trend in water depth at many locations.

Iron formations occur in Archean greenstone belts as well as in some late Precambrian and Phanerozoic sequences. Such deposits, however, are small compared to those formed during the early Proterozoic. The Archean iron formations and those younger than about 1800 million years are intimately associated with volcanic rocks, which probably provided a direct source of the iron and silica. They are referred to as "Algoma-type" iron formations and consist mainly of iron-rich clastic rocks and interbedded cherts that probably represent periods of reduced volcanic activity.

Most of the world's iron formations were deposited between about 2600 and 1800 million years ago. These "Superior-type" iron formations have puzzled geologists for a long time. Most are not associated with volcanic rocks and clastics are conspicuously absent. The sediment was formed more or less in place and so the iron and silica that are the dominant components of the rock

FIG. 3-18 A typical banded iron formation, early Proterozoic, Quebec. The iron is concentrated in the dark layers. (Canadian Geological Survey)

must have been transported in solution to the site of deposition and precipitated directly from the water. This process is not possible on a large scale today, for iron released by weathering is precipitated in soils at the source, where it is oxidized by contact with oxygenated surface waters and the atmosphere. Silica does not precipitate today (except in certain special environments, such as hot springs) because silica-secreting plankton (diatoms, silico-flagellates, radiolarians) keep both fresh and marine waters strongly undersaturated by drawing silica from the water to build their skeletons. In the Precambrian, however, there were no silica-secreting planktonic organisms to keep waters undersaturated with silica.

Various depositional environments have been proposed for the banded iron formations, including hot springs, hypersaline playa lakes, and the ocean, but most workers seem to favor a marine environment. Sedimentary structures, types of grains, and grain size distributions are very similar to those of the modern shallow-water carbonate deposits of the Persian Gulf and Bahama Banks. The small-scale iron-poor and iron-rich layering has been explained in several ways. Free oxygen decomposes organic matter and many basic biological reactions cannot proceed in its presence. Some geologists believe that the layering in banded iron formations resulted from cyclic expansion of the biomass of photosynthesizing blue green algae, which relied on dissolved ferrous iron in the water to act as an oxygen sink to carry the oxygen away from the cells. Procaryotes, such as blue green algae, carry on their metabolic reactions on the outside of the cell wall. In an ocean having a significant amount of iron in solution, oxygen liberated by photosynthesis could have been rendered harmless by combining with the iron to form an insoluble precipitate that produced the iron-rich layers of the banded iron formations. It has been suggested that the photosynthesizers periodically expanded in numbers to such a degree that the oxygen-mediating capabilities of the seawater were exceeded. As their numbers dropped drastically in response to oxygen poisoning, an iron-poor layer was deposited. The succeeding expansion of the biomass, as iron concentrations in the water rose again, produced another iron-rich layer and so on.

Other geologists have related the layering of banded iron formations to some type of periodic event, such as seasonal changes in the productivity of oxygen-liberating photosynthesizers or in upwelling, which would increase the amount of iron available in surface waters. All such interpretations require at least slightly evaporative conditions to create supersaturation with respect to silica. In any case, we would expect layering to be well preserved in Pecambrian rocks, whatever its origin, because there were no bottom-dwelling, burrowing, and sediment-ingesting organisms to churn the sediments after deposition.

The various depositional environments in which banded iron formations may have formed (for example, marine lagoonal, oölitic, and outer shelf or some nonmarine environments, such as hot springs and hypersaline lakes) suggest that the conditions that permitted the deposition of the iron and silica were characteristic of the atmosphere–hydrosphere system rather than some particular local depositional environment. Many geologists believe that the development of banded iron formations represents a time when photosynthesizers were present but oxygen concentrations were still low enough to permit significant quantities of dissolved iron in the oceans.

Although most geologists agree that the sediments of the banded iron formations were deposited in environments similar to modern carbonate banks and shelves, some do not accept the concept of a Proterozoic low-oxygen atmosphere. These geologists believe that the sedimentary particles in banded iron formations were not precipitated directly as iron and silica but were originally calcite and/or aragonite (as in modern carbonates) and have been altered to their present mineralogy by diagenesis.

The Ozone Shield

Photosynthesis by blue green algae releases oxygen as a waste product. When blue green algae first began to release oxygen in the Precambrian, the oxygen quickly combined with dissolved iron and other oxygen-deficient elements and compounds in the surrounding waters rather than being released to the atmosphere. Eventually, when most exposed crustal rocks were oxidized, free oxygen began to accumulate slowly in the oceans and then in the atmosphere.

When the atmosphere's oxygen content reached about 1 percent of its present level, a layer rich in ozone formed high in the atmosphere. Ozone (O_3) is still produced there today by intense solar radiation at an elevation of about 25 kilometers and is extremely important to us because it shields the earth's surface from lethal ultraviolet radiation. Prior to the existence of this layer, life could not exist on land or even in the upper, shallowest portion of the ocean. After the ozone layer formed, organisms were able to occupy the upper layers of the oceans, including the vast areas of shallow water fringing the continents. The plentiful solar energy available to these environments stimulated organic productivity, which, in turn, accelerated the accumulation of free oxygen. The newly habitable sunlit surfaces of the earth awaited the first terrestrial organisms, which may have been algal-fungal colonies akin to present-day lichens. We do not know when an ozone layer was first achieved, but perhaps it was in the late Precambrian shortly before the appearance of diverse invertebrate groups at the beginning of the Paleozoic.

EARLY LIFE

Oxygen Metabolism and Evolution of the Metazoa

The appearance of abundant free oxygen made possible respiration and oxygen metabolism as we know it today. Oxygen metabolism is the way in which all higher organisms, animals and plants alike, break down compounds in order to produce energy. Many primitive bacteria, however, are anaerobic; that is, they derive their energy from chemical reactions that do not involve oxygen. The appearance of oxygen metabolism permitted a great increase in metabolic efficiency. Fermentation, for example, is a common kind of anaerobic energy-generating process used by many single-celled plants. The *fermentation* of glucose (a common type of sugar) produces 57 kilocalories of energy per mole. In contrast, the *oxidation* of glucose, as performed by all higher organisms, produces 637 kilocalories per mole, about 12 times more. The advantage to the organism is obvious. Oxygen metabolism also made possible the evolution of the eucaryotic cell, which, in turn, allowed the development of multicellular animals and plants. Afterward the way was paved for the vast surge of evolution of higher organisms in the Phanerozoic.

Today some tiny single-cell organisms, which are grouped with the bacteria, have the capacity to metabolize both ways. Normally they perform oxygen metabolism, but when deprived of oxygen, they revert to anaerobic metabolism. These organisms provide insight into the versatile organisms that must have evolved when free oxygen was first beginning to accumulate. The switch from oxygen to anaerobic metabolism and back again takes place at an oxygen concentration of about 1 percent of the present atmospheric level (PAL), which is probably about the concentration that needed to be achieved in the atmosphere before oxygen metabolism became possible. It is difficult to establish with certainty that particular Precambrian microfossils were indeed eucaryotic rather than procaryotic, but evidence provided by measurements of cell size and the possible discovery of eucaryotic cells preserved in the act of dividing suggests that eucaryotes evolved about 1400 million years ago.

Because oxidative metabolism seems to require at least 0.01 PAL oxygen concentrations, the appearance of eucaryotes at about 1400 million years ago may signal the achievement of this level of oxygen in the atmosphere. Additional evidence of rising oxygen levels before 1400 million years is also provided by the appearance of red beds at about 2000 million years and the disappearance of Superior-type banded iron formations, which probably required the transport of reduced iron in solution in the ocean, around 1800 million years ago. The increase in oxygen content to its present-day level may have occurred gradually throughout geologic time or may have been erratic, particularly if overall production of marine plants, the primary oxygen source, varied significantly over the long term.

The First Fossils

The oldest convincing fossil remains on earth come from chert deposits in the 3200-million-year-old Fig Tree Series of South Africa. Thin sections cut from these chert beds reveal microscopic filamentous and circular cells of bacteria and blue green algae. Similar fossils have been found in a few other very ancient rocks, such as the Gunflint Chert in the Lake Superior region of North America (Fig. 3-19).

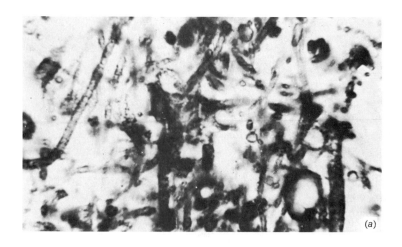

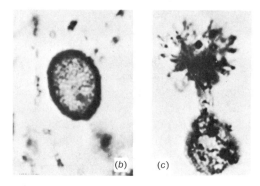

FIG. 3-19 Procaryote remains from the 1.9-billion-year-old Gunflint Chert of Ontario, Canada (magnification 1000 to 2000 times): (a) Thread-shaped forms that closely resemble modern filamentous bacteria and blue green algae. (b) Spherical form resembling a modern bacterium. (c) Parachute-shaped form of unknown affinity. (Courtesy Elso S. Barghoorn)

Precambrian strata throughout the world do yield one kind of fossil in considerable abundance—the stromatolites, which are considered to be fossils in the same sense as tracks or burrows in younger rocks. Although not representing primary organic remains, they are distinctive configurations of carbonate laminae produced by blue green algae (Fig. 3-20).

Stromatolites are being deposited today in intertidal carbonate environments and in supratidal environments, which are wetted only occasionally during storms or unusually high tides. In these environments the sediment surface is coated with a layer of cells of blue green algae. The algal cells are filamentous and bind the underlying and surrounding tiny particles of carbonate material into a coherent layer called an "algal mat." When the area is flooded during a storm or a high tide, a thin layer of tiny calcium carbonate particles is trapped by the numerous microscopic filaments on top of the algal mat.

FIG. 3-20 Precambrian algal stromatolite about 2 billion years old. Nash Fork Formation, Medicine Bow Mountains, southern Wyoming. Thickness of the bed is about 40 centimeters.

Quickly, within hours, the algae extend their filaments through this fresh layer, thereby incorporating it into the mat as they repopulate the new top surface. With the next influx of carbonate-laden water, the process repeats, producing the characteristic thinly laminated structure. The algal layers themselves are not preserved but are represented only by the bedding planes between the laminae.

Algal stromatolites occur in a variety of forms—flat, wavy, moundlike "heads," and even in columnar forms. Today columnar stromatolites are rare, but in the late Precambrian many types occurred, some of which appear to have had comparatively short ranges in time. Russian geologists, in particular, have found them useful in correlation and have divided the upper half of the Proterozoic into four zones based on stromatolite genera. Other workers believe that much of the morphologic distinction in stromatolites merely reflects differing local environments. It is well known, for example, that columnar stromatolites today and in ancient strata form elongate heads whose overall shape is governed by the direction and strength of local water currents (Fig. 3-21). Yet the tiny detailed structures of stromatolites seem to be biologically (genetically) controlled. This situation is best shown in well-documented Precambrian occurrences where the same types of stromatolites are found in several different kinds of rocks that reflect widely differing local environments. The implication is that while gross shapes of the stromatolite algal colonies may be environmentally controlled, most of the numerous detailed structures within the growing heads appear to be genetic and to represent truly distinctive taxa.

FIG. 3-21 Current directions were parallel to the elongate shape of these Precambrian stromatolites. (Courtesy Paul Hoffman, Geological Survey of Canada)

Biotic Change and the End of Precambrian Time

The most dramatic and extensive evolutionary radiation in all of earth history took place around 600 million years ago and it serves to separate the Precambrian and Phanerozoic intervals of earth history. All but a few of the principal kinds of life found on the earth today originated during this period.

The oldest fossil invertebrates occur as a peculiar assemblage of about six kinds of distinctive soft-bodied forms that exist as impressions in sandstones. They are found in rocks of about the same age in several parts of the world. The best-preserved examples are found in the Ediacara Formation of southern Australia and include animals resembling present-day jellyfish, sea pens, and segmented worms (Fig. 3-22). The first occurrence of this "Ediacara fauna," as it is called, appears to be around 650 to 700 million years old. Some of the organisms that made up the Ediacara fauna achieved fairly large size and it is probably no accident that they were very thin. The oxygen requirements of single cells surrounded by water can easily be met even at low oxygen concentrations by diffusion across the cell membrane. Aggregates of cells, as in the metazoans that formed the most conspicuous part of the biota since the Cambrian, require higher levels of oxygen and most have developed specialized respiratory systems to deliver the oxygen to cells buried within their tissues,

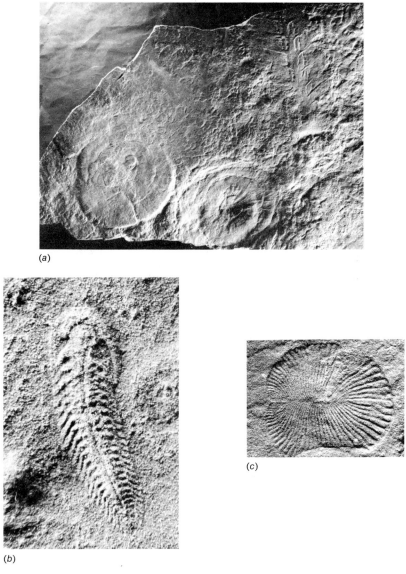

(a)

(b)

(c)

FIG. 3-22 The oldest known fossil animals from the late Precambrian Ediacara Hills of south Australia. The fossils are preserved as impressions in sandstone. (a) Rounded, jelly-fishlike forms and a branched sea penlike form (upper right) (one-third natural size). (b) Elongate, wormlike form (magnified two times). (c) Rounded, wormlike form (actual size). (Courtesy M.F. Glaessner)

where oxygen cannot reach them by diffusion. The thin, planar shapes of most members of the Ediacara fauna maximized their surface-area-to-volume ratio and facilitated the diffusion of oxygen into relatively large metazoans that probably had no specialized respiratory systems.

Studies of the oxygen requirements of modern eucaryotes in low oxygen environments give us some idea of the oxygen levels that may be represented by the appearance of the Ediacara fauna in the late Proterozoic and of skeletal fossils in the Cambrian. In modern low oxygen basins bottom-dwelling invertebrates are present where oxygen is less than 0.1 milliliter per liter (equivalent to 0.01 PAL in the atmosphere). Small, soft-bodied metazoans can exist above about 0.3 to 0.6 milliliters per liter and above 1.0 milliliter per liter (10 percent PAL) a diverse calcareous fauna exists. Thus the Ediacara fauna may represent the achievement of about 6 percent PAL.

In Phanerozoic sedimentary rocks fossil impressions of soft-bodied animals, such as those in the Ediacara fauna, are rare compared to the remains of *shell-bearing* animals, which, of course, are much more readily fossilized. About 570 million years ago, relatively soon after the appearance of the Ediacara fauna, many kinds of shell-bearing invertebrate animals became abundant and their remains occur in marine sedimentary rocks of this age all over the earth. It is the appearance of these abundant fossil shells that marks the traditional beginning of Cambrian time.

Rather surprisingly, the most abundant animal fossils in Cambrian rocks are not small, simple, one-celled forms as might be expected. Instead we find many-legged *trilobites*, which are distant relatives of modern crabs, and *brachiopods*, two-shelled, superficially clamlike animals. The abundance of these relatively advanced animals indicates that the initial radiation of metazoan life rapidly led to complex structures and adaptations.

The puzzling question concerns the reason why varied and abundant animals appeared in the Cambrian Period whereas prokaryotic plants, from which the animals probably arose, had existed through most of the preceding 3400 million years of Precambrian time. Although the cause is still obscure, this time marked the beginning of numerous lineages of rapidly evolving organisms that left a striking record of biotic change throughout the remainder of the Phanerozoic. Was this truly an abrupt evolutionary proliferation of metazoa (multicelled animals) or did the sudden profusion of fossils result simply from the development of preservable hard skeletons by animal groups that had existed for a long time?

We do not know the answer. Lately, however, it has become more and more difficult to defend the idea that soft-bodied metazoan animals might actually have been common through much of late Precambrian time but were not preserved as fossils because of their lack of shells or skeletons. It is now known that even soft-bodied metazoans leave distinctive patterns of trails and burrows in the ocean-floor muds and sands in which they live. Recent studies of such animal-produced structures in Precambrian sedimentary rocks show that they appear in considerable abundance about 700 million years ago, at about the time of the Ediacara fossils. The Ediacara fauna itself is made up entirely of soft-bodied jellyfishlike and wormlike animals preserved as mere impressions in sandstone. Although such preservation requires rather special environmental conditions, the Ediacara fauna has been found in rocks of very late Precambrian age on several continents. Thus the lack of unequivocal metazoan trace fossils in older Precambrian rocks strongly suggests that metazoa did not yet exist and that the abrupt appearance of invertebrate fossils in the early Paleozoic is truly a part of the burst of metazoan evolution that began in the latest Precambrian. The reason for this evolutionary burst remains a fundamental puzzle of life history.

Many suggestions have been advanced to explain this puzzle, most of which fall into two broad categories. The first relates the rapid metazoan expansion to direct changes in the organization of the animals themselves rather than to changes in the physical environment. The second relates the rapid metazoan expansion to the removal or surmounting of some ecological barrier that had previously prevented the diversification of metazoan life. Consider briefly an example of each of these kinds of theories.

One idea, which involves fundamental changes in the physical and physiological organization of the metazoa, attributes the rise of multicellular animals (and multicellular plants as well) to the development of sexual reproduction. Throughout most of geologic history, simple asexual division was the way all organisms reproduced. Even today single-celled organisms can quite easily reproduce this way. First, critical components are duplicated within the cell and then the cell simply separates into two individuals. The vast majority of multicellular organisms, however, reproduce sexually; here germ cells from different individuals unite to form an offspring. Sexual reproduction offers enormously greater prospects for modifying the genetic material of the offspring — and hence its structure and adaptation — than simple asexual fission. For this reason, some workers have suggested that the rapid expansion of multicellular animals and plants, beginning in the late Precambrian, simply marks the time when sexual reproduction first appeared.

Other geologists have suggested that low levels of atmospheric oxygen prevented the diversification of the metazoa. According to this hypothesis, the appearance of the soft-bodied Ediacara fauna, followed (about 100 million years later) by the appearance of abundant and relatively diverse skeletalized forms represents the achievement of 6 percent and then 10 percent, respectively, of the present atmospheric oxygen level. In this view, skeletons appeared when oxygen levels were high enough to support the extra energy required to build a skeleton and to permit the concomitant loss of surface area exposed to the water. Skeletons have several advantages. They provide protection from predators and unfavorable conditions, such as exposure at low tide, high sediment influx, and salinity fluctuations. They also provide rigidity and attachment for muscles, which greatly facilitates locomotion.

These speculations about the oxygen levels in the late Precambrian and early Paleozoic atmosphere are appealing but do not prove what the atmospheric oxygen levels actually were. The long-term production of oxygen by photosynthesizers since at least 3 billion years ago may have oxidized all the crustal rocks and then provided a much higher oxygen level than 10 percent PAL substantially before the Cambrian. In this case, we would need to look elsewhere for the evolutionary impetus, or perhaps hindrance, that was responsible for the actual timing of the observed events in the fossil record.

four

the paleozoic earth

The approximately 570 million years that have elapsed since the close of the Precambrian are commonly referred to as the *Phanerozoic Eon* (literally the "evident-life eon"), a name that reflects the importance of fossil shells and skeletal material in strata of post-Precambrian age. These fossils provide not only a revealing record of life but also the tools for a much more reliable correlation of strata and interpretation of sedimentary environments than can be achieved for rocks deposited previously. As a result, we have a much clearer understanding of Phanerozoic than Precambrian history. The Phanerozoic is divided into three eras: Paleozoic, Mesozoic, and Cenozoic (for ancient, middle-, and modern-life). The first 345 million years, approximately three-fifths of Phanerozoic time, belong to the Paleozoic Era (Fig. 4-1).

PALEOZOIC CONTINENTS AND OCEAN BASINS

Epicontinental Seas

Although shallow seas had sporadically occupied the interior portions of the continents at least since the close of the Archean, Paleozoic rocks contain the first clear-cut records of extremely widespread advances of the sea into continental interiors. If similar large-scale marine floodings of the continents occurred in the Precambrian, they are not discernible because much of the Precambrian stratigraphic record was removed during a widespread episode of erosion in latest Precambrian time. During this episode continents stood extraordinarily high above sea level, much as they do today. The erosion surface produced at this time is overlain by Cambrian sedimentary rocks that record a

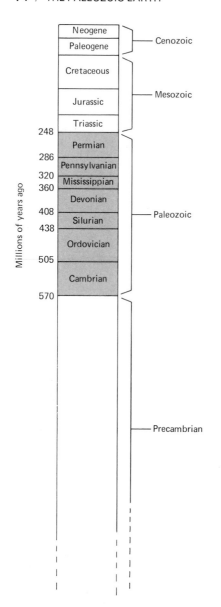

FIG. 4-1 The Paleozoic interval of earth history.

long, slow transgression of Cambrian seas onto all of the continents. In most regions the contact between the Cambrian and the underlying Precambrian is easily recognized because it is a pronounced unconformity. Along some Cambrian continental margins, as in the Great Basin region of the United States, however, deposition of nearshore sediments actually began in the late Precambrian and continued throughout much of the Cambrian. In such regions the Precambrian–Cambrian contact commonly lies within thick sequences of sandstones and the only way to locate the boundary is through fossils. Doing so is difficult because this boundary is marked not by a fossil *change* but by the lowest *appearance* of shelled invertebrates. Because many basal Cambrian strata are not fossiliferous, this contact in many areas is picked with considerable uncertainty.

The great marine transgression in the Cambrian began a pattern of continent-wide invasions and retreats of the sea that continued throughout the Paleozoic. Sometimes large portions of the continents subsided below sea level and became almost totally submerged by shallow, widespread epicontinental seas (Fig. 4-2). At other times, the lands stood high and the marine waters lapped only onto the edges of the continental shelves. Even while largely submerged, the continents remained very real entities. They did not subside to oceanic depths but instead stood as high platforms above the surrounding ocean basins.

The widespread epicontinental seas that formed during the Cambrian persisted through the Early Ordovician and then retreated, exposing vast regions. In Middle Ordovician time the seas again began a slow invasion that culminated in the most widespread epicontinental seas of which we have a record. But again these seas retreated, beginning in Middle Silurian time, completing the second major transgressive-regressive cycle. The next large-scale invasion of the sea began in the Devonian, but the seas again retreated late in

FIG. 4-2 Kinds of sedimentary rock formed in the Late Cambrian sea in North America. Coarsest detritus is adjacent to cratonic source area.

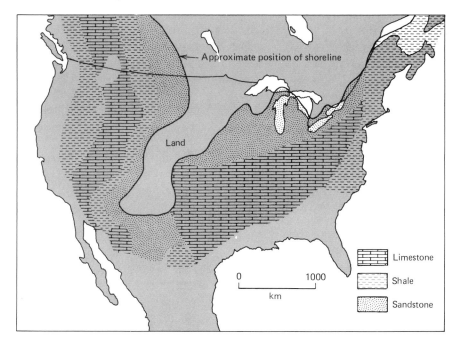

the period. Additional minor advances of the sea continued throughout the remainder of the Paleozoic, but more and more of the land surface became exposed and nonmarine sediments accumulated over progressively wider areas until by Late Permian time most parts of the continental platforms stood above sea level.

The cause of these large-scale marine transgressions and regressions is unknown. Either the continents rise at times and subside at others or the ocean basins of the world periodically decrease in capacity and overflow. The latter suggestion is favored by the coincidence on several continents of the transgressive-regressive episodes. This subject is discussed further in Chapter 5.

Paleozoic Plate Tectonics: the Assembly of Pangaea

At the beginning of the Paleozoic the shapes and the arrangement of the continents were quite different from today's familiar geography. Then, throughout the Paleozoic, these continents converged one by one and joined as the oceanic crust between them was subducted. By the end of the Permian, the contents had been assembled into one huge supercontinent called *Pangaea.*

We have a far better idea of how the continents have been distributed on the earth's surface since the Triassic — when Pangaea broke apart to produce the modern continents — than for the Paleozoic. The outlines of modern continents cannot help to reconstruct the pre-Triassic continents because the modern continental shapes all originated with the post-Triassic breakup. Furthermore, we cannot reconstruct positions of the continents based on the ages of intervening ocean floor because none of the present ocean floor was formed before the Triassic; all the ocean floor that existed in the Paleozoic has been consumed by subduction. Consequently, we must rely mainly on two methods for tracing the positions of the Paleozoic continents: (1) the positions of ancient magnetic poles, based on rock magnetism, and (2) the distribution of Paleozoic geosynclinal foldbelts, which can commonly be interpreted as the zone of collision between two ancient continental margins. An additional tool, the reconstruction of ancient biogeographic provinces, has been of corroborative value in some cases.

Although we can determine the ancient orientation of a continent in relation to the North and South poles and the latitude of the continent, we cannot determine the longitude. This means that, from paleomagnetic data alone, we cannot tell how far the continents have moved longitudinally in the geologic past. Typically both latitudinal and longitudinal components are involved in a continent's motion. The latitudinal component can be resolved readily; the longitudinal component, however, is always elusive.

Figure 4-3 shows that several Paleozoic mobile belts sliced across Laurasia, the northern hemisphere portion of the supercontinent. All these mobile belts contain ophiolites of Paleozoic age and thick, highly deformed sequences of Paleozoic sedimentary rocks. The ophiolites are inferred to be rem-

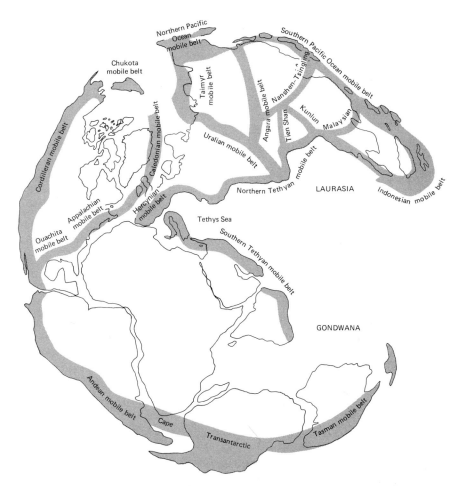

FIG. 4-3 The Triassic supercontinent of Pangaea, showing its two major subcontinents. Foldbelts younger than 450 million years border the Gondwana supercontinent, but none crosses it, suggesting that Gondwana had been intact at least since the early part of the Paleozoic Era. The Paleozoic mobile belts show how Pangaea was assembled. (Seyfert and Sirkin, 1973)

nants of the ocean basins that once lay between the continental blocks. Thus it appears that prior to the construction of Pangaea the continental blocks between the mobile belts were individual entities and may have been widely separated. No mobile belts cut across Gondwana, the southern hemisphere portion of the supercontinent, although they did form a nearly continuous belt around it. Thus Gondwana appears to have existed as a single continent throughout the Paleozoic Era whereas Laurasia is a mosaic of several continental blocks assembled during the Paleozoic.

Figure 4-4 shows how the positions of the Paleozoic continents changed between Cambrian and Devonian time. The precise longitudinal spacing of

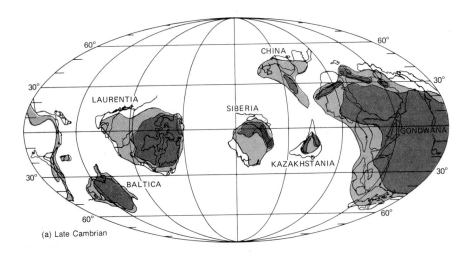

(a) Late Cambrian

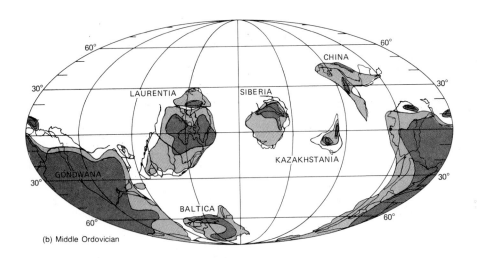

(b) Middle Ordovician

FIG. 4-4 Early Paleozoic paleogeography: (a) Late Cambrian; (b) Middle Ordovician; (c) Middle Silurian; (d) Early Devonian. Mollweide projection showing entire earth surface. (Scotese and others, 1979)

these early Paleozoic continents is tentative, but their relative positions and latitudes are essentially correct. During the Cambrian and the early part of the Ordovician the ancestral North American (Laurentia) and European (Baltica) continents lay on opposite sides of a widening ocean. This ocean is commonly

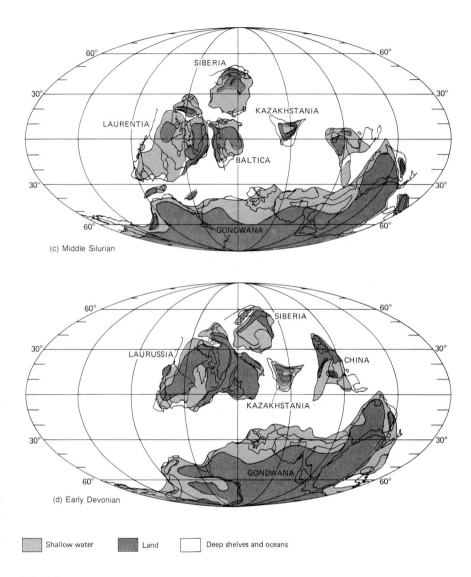

(c) Middle Silurian

(d) Early Devonian

| | Shallow water | | Land | | Deep shelves and oceans |

FIG. 4-4 continued.

referred to as the Iapetus Ocean, and although it separated much of what is now Europe from much of what is now North America, it is not the ancestor of the modern Atlantic.

Paleomagnetism and structural and faunal evidence indicate that parts of present-day New England and Canada's maritime provinces were part of

Baltica during the Paleozoic and much of the southeastern United States was part of Gondwana. Scotland and Northern Ireland were apparently part of Laurentia (Fig. 4-5). In the Middle Ordovician large-scale thrust faulting took place in the northern Appalachians of the United States and Canada. This earliest mountain-building event along the eastern margin of Laurentia is named the *Taconic Orogeny* from the Taconic Mountains of easternmost New York State. Somewhat later, in Late Ordovician time, a highland arose along the former continental margin in the same region and shed a large volume of detrital sediments westward onto the former continental shelf. Today in New York and Pennsylvania, Upper Ordovician shallow marine shales and limestones may be traced eastward into nonmarine red beds, chiefly sandstones and shales. The entire sequence represents a delta that built from the east into the shallow epeiric sea. It has been called the *Queenston Delta*.

In the northeastern United States outcrops of Ordovician volcanic rocks indicate the initiation of subduction and the formation of a volcanic arc along the margin of the continent. The Iapetus Ocean had begun to close (Fig. 4-6). In both the Canadian maritime provinces and northern Europe volcanism continued well into the Silurian. In the Late Silurian the geosynclinal strata in Britain and Scandinavia were strongly folded in an episode called the *Caledonian Orogeny.*

Rising highlands at the eastern margin of North America in Middle and Late Devonian time again shed large volumes of sediment westward onto the continent. Simultaneously, rising highlands shed sediments eastward onto Britain and Scandinavia. The resulting sedimentary rocks consist in part of thick sequences of fluvial red beds, including the Old Red Sandstone in Britain and

FIG. 4-5 In early Paleozoic time part of New England and Canada's maritime provinces belonged to the European continent and part of Great Britain, Ireland, Norway, and Spitzbergen belonged to the North American continent. During the Caledonian Orogeny the continents were joined. Much later, during the Mesozoic, they separated along slightly different boundaries. (Wilson, 1966) Reprinted by permission from *Nature*, vol. 211, pp. 677-78. Copyright © Macmillan Journals Limited.

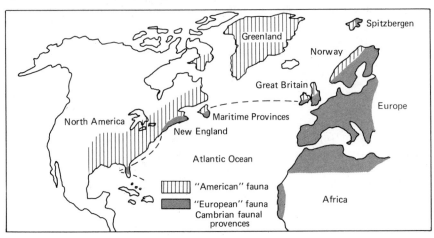

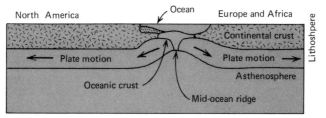

Late Precambrian: Iapetus ocean first opens

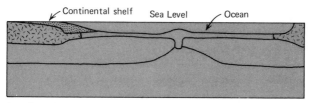

Late Cambrian: Iapetus ocean reaches maximum size.

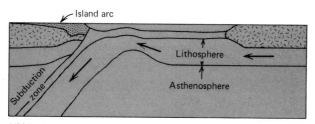

Middle Ordovician: Iapetus ocean begins to close. Crust is subducted beneath American continental margin.

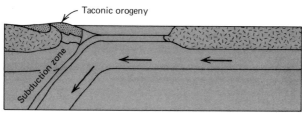

Late Ordovician-Early Silurian: Subduction continues. Island arcs and eastern margin of North America deform

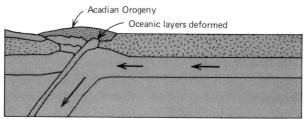

Late Devonian: Iapetus ocean vanishes as continents collide during the Acadian Orogeny further elevating northern Appalachians

FIG. 4-6 One idea of how the Iapetus Ocean opened and closed in the early Paleozoic.

the Catskill Formation in New York, Pennsylvania, and West Virginia. The similarity of rocks and faunas in both North America and Europe suggests that they were derived from opposite sides of a single elongate mountain belt that formed as Laurentia and Baltica finally collided. In New York and Pennsylvania the thick, fluvial red beds of the Catskill Formation can be traced westward into nearshore marine sandstone that, in turn, grades into offshore shales and limestones (Fig. 4-7). These deposits, which are called the *Catskill Delta*, attain a thickness of 3000 meters (Fig. 4-8) and they built westward for 500 kilometers behind the regressing Late Devonian epicontinental sea. The deposits of the Catskill Delta are similar to those formed earlier by the Ordovician Queenston Delta, but the Catskill deposits are thicker and more widespread, reflecting a higher and most persistent source area to the east.

Regional metamorphism and granitic intrusives of Devonian age in New England and the adjacent maritime provinces of Canada, testify to the intensity of the mountain-building activity that formed the highland source area. The tectonic episode in this region has been named the *Acadian Orogeny*. The suture zone between Laurentia and Baltica continued to be the site of tectonic activity long after the collision. It was the site of igneous activity in the Mississippian and, with periodic renewed uplift, it served intermittently as a source area for large quantities of detrital sediment during the late Paleozoic.

Thus the Taconic, Caledonian, and Acadian orogenies were simply regional expressions of the same process: the closing of the Iapetus Ocean and the suturing of Baltica to Laurentia. And this was only the first step in the uniting of all the continents to form Pangaea. Throughout the remainder of the Paleozoic, the separate continental blocks continued to converge on one

FIG. 4-7 Middle and Upper Devonian strata in southern New York coarsen eastward toward the highland source area formed when the ancestral North American and ancestral European continents collided during the Acadian Orogeny. This is a reconstruction of the strata at the end of the Devonian. The line of section is shown in Fig. 4-8.

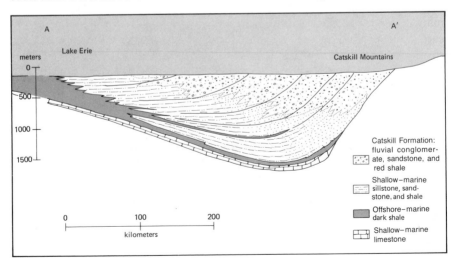

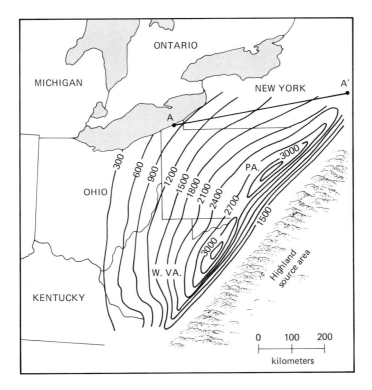

FIG. 4-8 Isopach map of fluvial and shallow-marine sandstones and shales deposited in the Catskill Delta during the Middle and Late Devonian. A-A' is the position of the cross section in Fig. 4-7. Isopachs in meters.

another. Their paths are documented by a growing body of paleomagnetic data and their actual collisions are recorded by several Paleozoic foldbelts.

The Ouachita geosyncline in Oklahoma and Arkansas lay at the southern margin of the Paleozoic North American continent [Fig. 4-9(a)]. During its early history this area quietly accumulated several thousand meters of limestone, sandstone, and shale and there is no evidence of volcanism. [Fig. 4-9(b)]. This situation probably indicates that the southern margin of the Paleozoic North American continent was not bordered by a subduction zone. Finally, in Mississippian time the Ouachita geosyncline began to receive sediments from a volcanic source area that lay to the south [Fig. 4-9(c)]. Increasing quantities of volcanic tuff in the Mississippian shales suggest an encroaching volcanic arc and subduction zone that bordered an approaching continent.

In the Pennsylvanian boulders derived from orogenic uplift of the outer part of the Ouachita geosyncline were transported northward by submarine slumps and debris flows along the steep scarps produced by the northward thrust faulting of huge crustal slabs that accompanied the encroaching subduction front [Fig. 4-9(d)]. In the Late Pennsylvanian and Early Permian thrusting and folding continued as a result of regional compression and the Ouachita

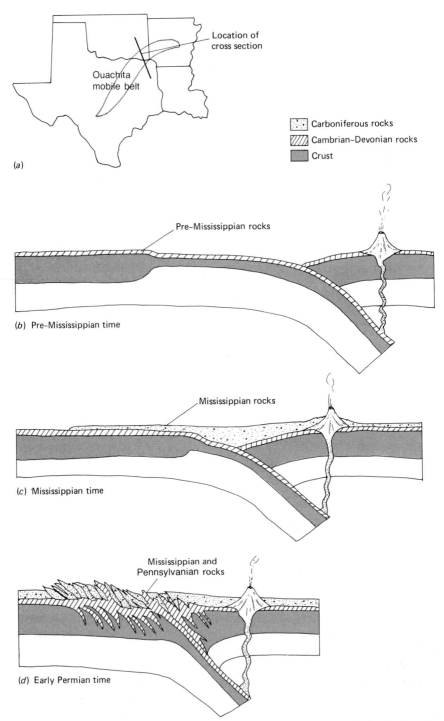

Location of
cross section

Ouachita
mobile belt

(a)

:::: Carboniferous rocks

/// Cambrian–Devonian rocks

▓ Crust

Pre–Mississippian rocks

(b) Pre–Mississippian time

Mississippian rocks

(c) Mississippian time

Mississippian and
Pennsylvanian rocks

(d) Early Permian time

FIG. 4-9 As Gondwana approached southern North America, deposition in the Ouachita geosyncline reflected increasing volcanic activity. Final collision created the Ouachita Mountains in the late Paleozoic. (Wickham, Roeder, and Briggs, 1976)

geosyncline became a fold-mountain chain, as this portion of ancestral North America collided with what is now South America.

The Marathon, Ouachita, and Appalachian geosynclines formed the southern margin of North America in the Mississippian (Fig. 4-10). Some of the modern portion of North America that lies south and east of these areas simply had not yet formed; much of what is now the southeastern United States was still part of Gondwana. Together with the Hercynian geosyncline of

FIG. 4-10 The Marathon, Ouachita, and Appalachian mobile belts in North America.

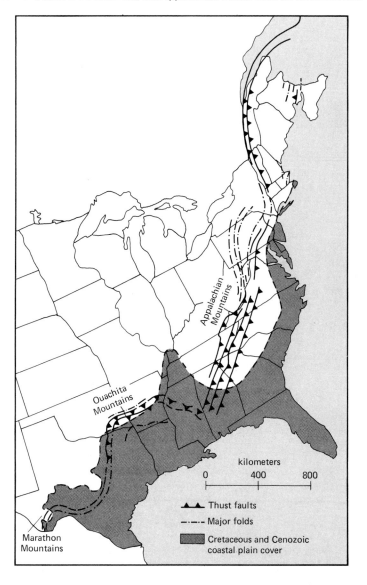

southern Europe (see Fig. 4-3), the Ouachita, Marathon, and Appalachian geosynclines formed an elongate belt marking the southern boundary of the continent of Laurussia, which included both the ancestral European and North American continents [Fig. 4-11(a)]. Beginning in the Pennsylvanian, the entire southern margin of Laurussia was strongly folded and thrust faulted as a result of collision with Gondwana [Fig. 4-11(b)]. This widespread late Paleo-

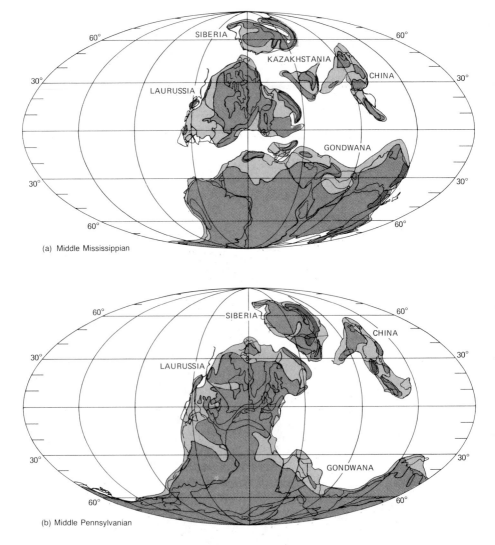

FIG. 4-11 Late Paleozoic paleogeography: (a) Middle Mississippian; (b) Middle Pennsylvanian; (c) Late Permian. Mollweide projection showing entire earth surface. (Scotese and others, 1979)

zoic tectonic activity is called the *Appalachian Orogeny* in the eastern United States and is known as the *Hercynian Orogeny* in much of the rest of the world. The Hercynian Orogeny represents the joining of the northern hemisphere continents with Gondwana to form the supercontinent of Pangaea.

During the late Paleozoic numerous geosynclinal belts in Asia were also deformed and uplifted into mountain ranges. By the Mississippian, at least four separate continental blocks, which now form Asia and a part of Europe, lay close together in the northern hemisphere. In addition to Laurussia, they included Siberia, Kazakhstania, and China. In the Pennsylvanian Kazakhstania joined Siberia. In Early Permian time Siberia and the attached Kazakhstan continental block moved southward to join Laurussia [Fig. 4-11(c)]. The collision created the Ural Mountains, which today divide Europe from Asia. Meanwhile, China moved westward toward the remainder of the world's continents to complete the supercontinent of Pangaea.

During most of Phanerozoic time the western margin of North America was the site of subduction zones and associated volcanic arcs. During the Paleozoic and Mesozoic the continental margin lay well inland of its present position, which was achieved only in the late Cenozoic. In the early Paleozoic sandstones and carbonate rocks accumulated on a tectonically quiescent shelf

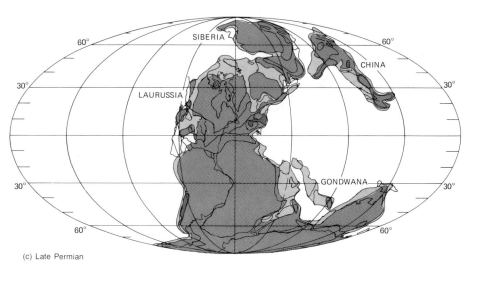

(c) Late Permian

Shallow water

Land

Deep shelves and oceans

FIG. 4-11 continued.

that stretched from Alaska to California. West of the shelf lay a volcanic arc. From the Ordovician to the Early Devonian extensional back-arc basins developed between the arc and the continental margin [Fig. 4-12(a)]. Graptolitic shales and cherts were deposited in these basins.

In the Devonian and Early Mississippian compression at the continental margin resulted in the closing or narrowing of the back-arc basin and the obduction (overthrusting) of basinal sediments onto the continental slope and shelf along the Roberts Mountain Thrust in Nevada [Fig. 4-12(b)]. This event is called the *Antler Orogeny*. The shedding of large quantities of clastic detritus both east and west from the resulting highland marks the first time in the Phanerozoic that sediment was shed eastward onto the craton as well as westward toward the continental margin.

In the Pennsylvanian and Permian new back-arc basins developed that were floored with oceanic crust. The Late Permian and Early Triassic had renewed orogenic activity similar to that in the Devonian. At this time the back-arc basins again closed by thrusting of basinal sediments onto the continent (the Golconda Thrust Zone) and a part of the volcanic arc became accreted to the continent, where today it constitutes the Klamath Mountain block (Fig. 4-13). During this event, which is known as the *Sonoma Orogeny*, the continental margin was shifted considerably to the west.

PALEOZOIC ENVIRONMENTS

Source Areas for Detrital Sediments

At times of low stands of sea level the exposed interior portions of the Paleozoic continents were important sources of detrital sediments. These sources are reflected in sedimentary rock facies that coarsen toward the continental interior. A typical facies change, from limestones and shales formed well offshore to sandstones formed adjacent to the then-exposed interior source area, is shown in the Upper Cambrian facies map of Fig. 4-2. Other cratonic sources were created by local uplifts. During the Pennsylvanian Period, for example, a belt of uplifts and block mountains formed from southern Oklahoma to Idaho (Fig. 4-14). Some of these uplifts supplied large volumes of coarse detritus to adjacent basins.

Along zones where Paleozoic plates converged and continents ultimately collided, once quiet continental margins were transformed into active mobile belts and folding and uplift of these regions created important source areas for major episodes of sedimentation. Material eroded from these sources was shed onto adjacent continents where they typically accumulated on top of older deposits of the former inner continental shelf. The closing of the Iapetus Ocean beginning in Late Ordovician time, for instance, produced a highland

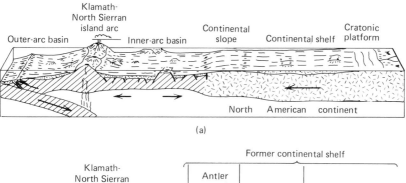

(a)

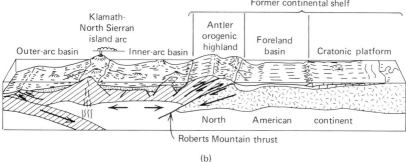

(b)

(c)

FIG. 4-12 A reconstruction of the middle Paleozoic paleogeography of the western United States: (a) Silurian to Devonian; (b) Latest Devonian to Mississippian. Obduction of back-arc basin crust onto the North American continental margin created the Antler orogenic highland, which shed large quantities of coarse detritus eastward into a foreland basin on the continent as well as westward into the back-arc basin. (c) Geographic location of the Antler Highland. The present position of the middle Paleozoic cratonic margin (as shown on the map) coincides in many places with the much later Mesozoic age Sevier Thrust. (Poole and others, 1977; Poole and Sandberg, 1977)

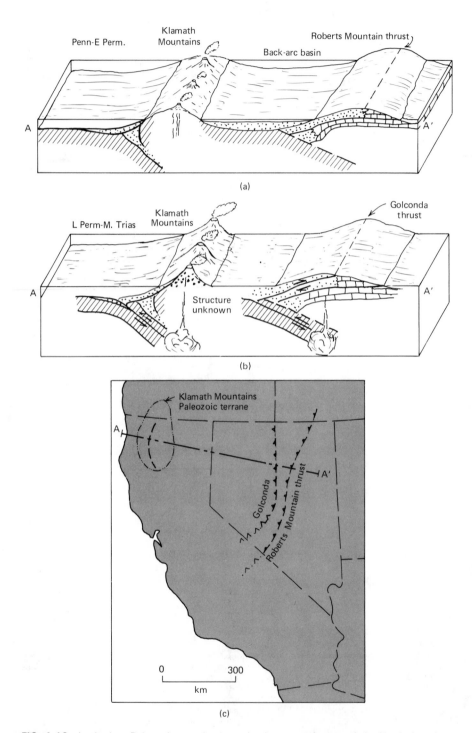

FIG. 4-13 In the late Paleozoic new back-arc basins opened west of the North American continental margin and these basins were closed as basin crust was obducted onto the continent along the Golconda Thrust. (Davis, 1974)

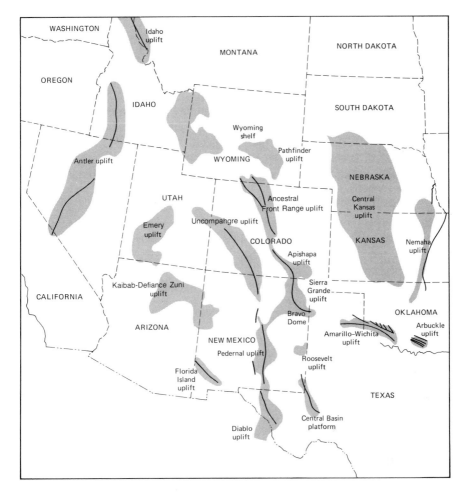

FIG. 4-14 Pennsylvanian uplifts in the western United States. Large quantities of coarse clastics were shed off some of these uplifts into adjacent sedimentary basins.

that shed coarse terrigenous sediments westward onto older Ordovician and Cambrian shelf carbonates and sandstones.

Carbonate Shelf Deposits

At those times during the Paleozoic when sea level stood high and continents were largely covered by shallow seas, potential source areas for terrigenous detritus were mostly submerged. Thick deposits of both limestone (calcium carbonate) and dolomite (calcium magnesium carbonate) accumulated over vast areas of the flooded continental platforms. During the Paleozoic little carbonate appears to have been deposited in the deep sea. Today, in contrast, most carbonate is being deposited in the deep sea as pelagic ooze. Shelf carbonates are forming in only a few places, but they provide valuable insights into their ancient counterparts.

The Bahama Banks: a Modern Model for Ancient Carbonate Deposition. Today, as in the past, the most important requisite for the accumulation of large quantities of carbonate sediments is that there be little influx of terrigenous detritus. In addition, most modern shallow-water carbonates are forming in low latitudes because far more carbonate-secreting organisms thrive in warm waters than in cold. The sand- and mud-sized particles that make up shallow-water carbonate sediments are derived chiefly from the breakdown of the skeletons of carbonate-secreting plants and animals, but inorganic precipitation of calcium carbonate directly from seawater also occurs in certain environments. Once produced, lime sands and lime muds are transported locally by waves and currents in the same way as terrigenous sands and muds, but compared to terrigenous sands and muds most carbonate grains do not move very far from where they formed. Carbonate sands show cross bedding and other structures indicative of high-energy water motion and carbonate muds tend to accumulate in a variety of sheltered, low-energy environments where wave and current motion is minimal. The Bahama Banks are perhaps the best-known area in which shallow-water carbonates are forming today and the study of this as well as other key areas has greatly enhanced our understanding of ancient carbonate shelf deposits.

Great Bahama Bank (Fig. 4-15) rises from the deep ocean floor 110 kilometers east of the Florida coast. The bank is a large, shallow platform almost 200 kilometers across but with water depths generally less than 5 meters. Tidal circulation over the interior of the bank is sluggish, with the result that water salinities and temperatures are generally higher than those of the surrounding deep oceans. The rocks exposed on the Great Bahama Bank, as well as the sediments being deposited, are exclusively carbonates. The carbonate sediments accumulating on Great Bahama Bank are chiefly muds, skeletal and oolitic sand, and "grapestone," which consists of moderately coarse aggregates of smaller grains that have been cemented together.

Lime mud consists, by definition, of the tiniest particles (less than one-eighth millimeter in diameter). On Great Bahama Bank most such particles are produced in the stems and branches of green algae. The mud accumulates in quiet-water environments of tidal flats and the adjacent subtidal environments of the inner portion of Great Bahama Bank. Here the mud contains occasional shells of mollusks and foraminifera (tiny shell-building protozoans), which live there. In the quietest environments it contains ellipsoidal mud pellets up to several millimeters in length, which are produced by burrowing mollusks and arthropods. Mud deposits also accumulate on tidal flats around the low islands. On the intertidal flats, which are flooded daily, blue green algae form temporary mats that trap mud particles. The mat-formed laminae are not preserved, however, because burrowing mollusks and arthropods repeatedly churn up the mud and homogenize it. Invertebrate shells are fairly common,

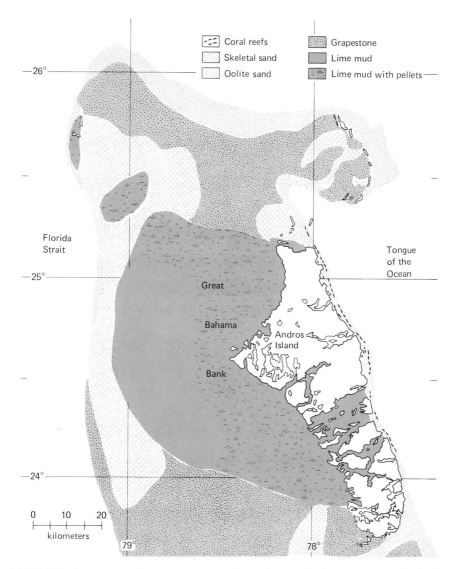

FIG. 4-15 Carbonate sediment types on the Great Bahama Bank are governed chiefly by salinity, current velocity, distance from the open ocean at the bank's margin, and distributions of carbonate-secreting organisms. (Laporte, 1968)

but their variety is limited. Few burrowing organisms live further landward on the supratidal flats, which are flooded only during storms or unusually high tides, and laminated muds trapped by algal mats are preserved to form stromatolites. Alternate wetting and drying in the supratidal environment produces mud cracks. Highly saline interstitial waters that periodically seep through the supratidal mats in many cases alter the previously deposited tiny grains of calcium carbonate to the mineral dolomite, $CaMg(CO_3)_2$. This kind of postdepositional alteration, referred to as "dolomitization," has been important throughout geologic history. In some regions, such as the Persian Gulf, where supratidal flats are extensive and the climate arid, evaporation of the occasional floods of seawater produces evaporite deposits, chiefly gypsum ($CaSO_4 \cdot 2 H_2O$) and halite ($NaCl$).

Carbonate sands reflect environments in which currents are generally too strong to permit mud to be deposited. Skeletal sand on Great Bahama Bank comes from the breakdown of calcium carbonate (chiefly the mineral aragonite) secreted by green algae, corals, mollusks, and other invertebrates that thrive mainly along the margins of the bank (Fig. 4-15). Sometimes the skeletal sand in this area is called "coralgal sand" for the two principal contributors of sedimentary particles.

Oolite sand consists of small spherical grains that are formed of concentric layers of calcium carbonate around a minute central nucleus. Oolite shoals accumulate in a belt 2 to 4 kilometers wide near the edge of the Great Bahama Bank where they are swept by tidal currents (Fig. 4-15). The concentrically-coated oolite grains form as they are rolled around in this high-energy environment by waters saturated in calcium carbonate. Few organisms live in this environment because of the unstable substrate. Oolitic limestones are common in the geologic record and evidence is good that they formed under similar environmental conditions.

The area on Figure 4-15 mapped as "grapestone" contains significant sand-sized grains that are cemented aggregates of fragmented smaller particles, chiefly oolite grains but also many foraminifera and various other particles. Grapestone apparently forms in areas where currents were formerly active but that are now only occasionally agitated, thus favoring cementation of the grains.

A Paleozoic Example. Limestone strata that appear to represent environments similar to those of Great Bahama Bank occur throughout the geologic column. During Late Cambrian and Early Ordovician time, for instance, extensive shallow-water carbonates formed on the continental shelf along the western margin of North America (Fig. 4-16). Simultaneously, deeper-water limestones formed on the continental slope immediately to the west. The shelf and slope limestones are well exposed in the Egan and Hot Creek ranges of eastern and central Nevada. Figure 4-16(b) is a reconstruction of their environmental setting, based on these exposures.

The shelf carbonates of the Egan Range are similar, in many ways, to those of the modern Bahama Banks. They occur as a vertical sequence of three major units. The lowermost unit is a thinly (1 to 2 centimeters) bedded, argillaceous (clayey), lime mudstone. The rocks are bioturbated; that is, they have been "stirred" by the activities of burrowing organisms. They contain trilobites, fecal pellets (peloids), and lenses of clay mudstone. The rocks of this unit formed in a low-energy, shallow, subtidal setting. The second unit, which overlies the first, consists mainly of stromatolitic algal laminae that form a

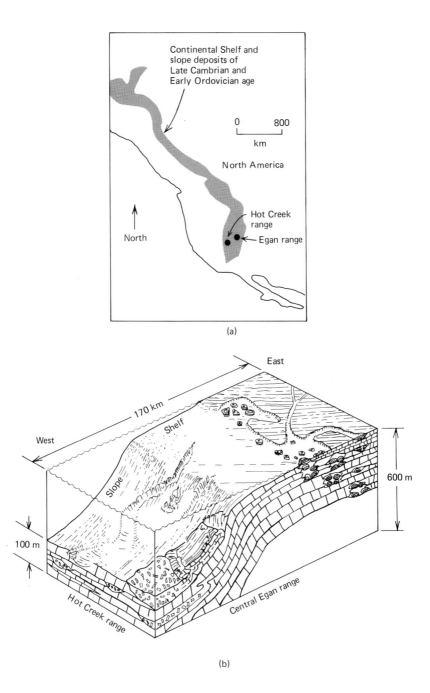

FIG. 4-16 (a) Distribution of Upper Cambrian and Lower Ordovician continental margin deposits of western North America, and location of the Egan and Hot Creek ranges of Nevada. (b) On the shelf, represented in the Egan Range, shallow-water limestones were deposited in algal buildups and on tidal flats. On the slope (represented in the Hot Creek Range), deep-water limestones formed and were redeposited, along with transported shallow-water sediments, by slumps, slides, and turbidity currents. (Cook and Taylor, 1977)

variety of shapes from laterally linked "heads" to sheets. The algal limestone interfingers locally with clastic limestones that consist of skeletal and algal grains and peloids. These grainstones were deposited as infilling between local buildups of the algae. This algal unit accumulated in a moderately high-energy, very shallow, subtidal-to-intertidal environment in which algal mats flourished. In the uppermost unit dolomitic lime mudstone and limestone containing large, irregular pore spaces alternate with flat-pebble conglomerate and breccia having clasts of the lime mudstone. Most of the conglomerate and breccia occur in shallow channels. Burrows and skeletal debris are rare. The presence of dolomite and limestone beds with large, irregular pore spaces generated by gas bubbles is a common feature of modern carbonate tidal flats and the overall environment of deposition for this uppermost unit was probably a tidal flat with adjacent tidal channels that were filled with storm deposits of rip-up clasts derived from the tidal flats. This vertical sequence of shelf-to-tidal flat carbonate units developed as the shelf prograded seaward.

At the same time that the shallow-water carbonates now exposed in the Egan Range were forming, deep-water limestones were being deposited 170 kilometers to the west on the continental slope. Some strata were deposited *in situ* (in place); they consist of dark, thin-bedded to laminated, fine-grained limestones containing deep-water trilobite species and sponge spicules. Redeposition was also common in this ancient slope area, as it is on modern continental slopes. The rocks of the Hot Creek Range provide evidence for a broad spectrum of mass movement processes in deep water. Some beds have undergone simple soft sediment deformation without shearing. Others have yielded and are both folded and broken. Still other units have thick (to 10 meters), massive conglomerates containing only clasts of deep-water origin or mixtures of deep- and shallow-water clasts. This conglomerate must have formed from subaqueous debris flows or turbidity currents.

The Modern Reef Model. Organic reefs consist of rigidly cemented skeletons of marine organisms that have accumulated in place. Reefs have been important locally in shallow tropical seas since the Early Cambrian. Today and throughout much of geologic history reef builders have been chiefly corals and carbonate-secreting algae. At times animals other than corals have dominated reef communities, but regardless of which were dominant at a given time, reefs appear to have always occupied a similar ecologic niche. Today reef organisms require a lot of sunlight and clear, agitated water. They flourish widely in low latitudes where the mean winter temperature of the water is warm — between 27°C and 29°C — and where salinity remains normal — that is, the same as that of the open sea. Reef organisms are closely packed and cemented together so that the remains of individuals do not readily break down to smaller pieces and erode away when they die. Instead, massive structures composed of lifeless carbonate skeletons persist indefinitely to form a foundation on which the living animals and plants of the reef constitute but a thin surface film. Reef organisms grow only on hard substrates where they can anchor themselves firmly. Their upward growth limit is the surface of the sea itself where modern reefs typically generate a surf zone. Reefs favor the windward side of shallow water platforms and islands in preference to the leeward side. Corals and algae build strong, firm buttresses that can withstand large waves, but they cannot survive in water clouded by suspended sediments.

The exact shape of a growing reef is molded by wind, waves, and currents, but most are fairly narrow — not more than few hundred meters across — and they generally form elongate barriers that grow in a high-energy wave zone between a shallow platform on one side and a comparatively deep marine basin on the other. The largest reef that is actively being built today is the Great Barrier Reef, which extends for 1600 kilometers (1000 miles) along the coast of northeastern Australia. At the other extreme, small, roughly circular *patch reefs* on tropical or subtropical shelves may be no more than a few meters in diameter. Elongate *barrier reefs* grow most actively on their ocean-facing *forereef* side (see Fig. 4-17). Here, where the reef maintains a steep pro-

FIG. 4-17 Reefs strongly affect regional sedimentation patterns. Each of the zones in a reef supports its own distinctive biota, which contributes characteristic sediments to the overall reef buildup.

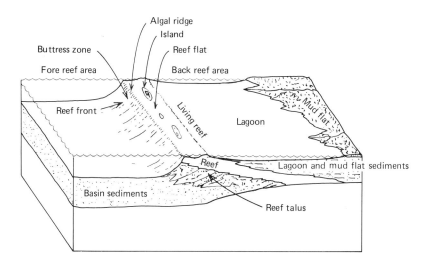

file called the *reef front*, any reef material that becomes loosened moves down-slope under the influence of gravity and forms a detrital apron of *reef talus*. The protected waters behind the reef on the landward or bankward side constitute a *back-reef* area. The water-covered portion of the relatively quiet, shallow back-reef area is called a lagoon. Typically it is very broad compared to the narrow reef zone itself. The back-reef area may be a suitable habitat for invertebrate organisms that cannot survive in the surf zone and that generate significant quantities of carbonate sand and silt. In arid climates, however, extensive back-reef areas may become important sites of evaporite sedimentation.

Paleozoic Reefs. The term *reef* is usually restricted to shallow-water, wave-resistant carbonate buildups supported by a rigid framework of skeletal material. However, the corals that provide the rigid framework of most modern wave-resistant reefs did not evolve until the Mesozoic and hence Paleozoic carbonate buildups that are generally called reefs were constructed differently. Both early and late in the Paleozoic these buildups lacked any rigid skeletal framework and were much like modern lime mud banks that form in comparatively low energy environments. Only from Late Ordovician to Devonian time were carbonate buildups dominated by corals that formed a skeletal framework, but even these corals were smaller than those of today. Although the Paleozoic reefs appear to have formed in somewhat deeper water and in lower energy environments than modern reefs, the ecological succession of their organisms was generally similar. In addition many reefs that formed throughout the Paleozoic, like modern reefs, strongly influenced depositional processes and hence the sedimentary facies around them.

Although the first appearance of skeletal fossils marked the beginning of the Cambrian, Cambrian faunas were not very diverse. Most fossils found in Cambrian rocks belong to two non-reef-forming groups: the trilobites and the inarticulate brachiopods. Both Cambrian and Early Ordovician reefs consisted mostly of lime mud that was trapped by algae. Many such lime mud mounds built upward into shallow water. Some Early Cambrian reefs contained significant numbers of Archaeocyathids, small cone-shaped skeletons of unknown affinity that superficially resemble corals.

Finally, in the Late Ordovician, a new diverse invertebrate fauna developed that included the rugose and tabulate corals, the bryozoans, and stromatoporoids, and representatives of several other groups, all of which became important contributors to Paleozoic reef communities (Fig. 4-18). Reefs that formed during the Late Ordovician, Silurian, and Devonian were dominated by stromatoporoids and corals both of which played an important role as frame builders.

In Ordovician and Silurian reefs tabulates were larger and more important than the rugosa (Fig. 4-19). Later, in the Devonian, the rugosa dominated. Neither of these groups had representatives that were as large as framebuilders in modern reefs. Most reefs that formed during this time began as a low mound with a lime mud core and the fauna was generally dominated by

FIG. 4-18 Distributions of organisms in and around a Devonian reef in Europe. This same pattern applies to nearly all post-Cambrian Paleozoic reefs and, with a few alterations, could be used to illustrate Mesozoic and Cenozoic reefs as well. (Krebs, 1974)

Organisms	Back-reef	Reef	Fore-reef	Basin
Massive & tabular stromatoporoids				
Upright branching stromatoporoids				
Tabulate corals				
Rugose corals				
Brachiopods				
Echinoderms				
Gastropods				
Cephalopods				
Bryozoans				
Ostracods				
Conodonts				
Algal coatings and crusts				

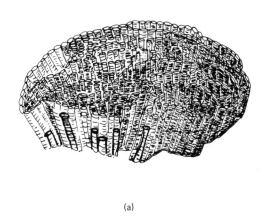

(a)

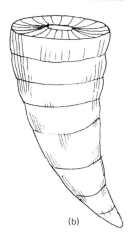

(b)

FIG. 4-19 (a) Tabulate and (b) rugose corals. Tabulates were the major coral frame builders in Late Ordovician and Silurian reefs; rugose corals were the major frame builders in Devonian reefs. These are shown slightly reduced.

bryozoans or sponges and algae (Fig. 4-20). These mounds often formed the base for larger pinnacle reefs, banks, and atolls, whose framework consisted mostly of rugose and tabulate corals. The large buildups, which probably grew nearly to the surface, were often capped by stromatoporoids. Debris from reef-dwelling organisms, particularly crinoids, accumulated on the flanks (Fig. 4-20). The Silurian pinnacle reefs of the Michigan Basin, which contain large

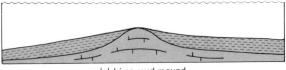

(a) Lime mud mound

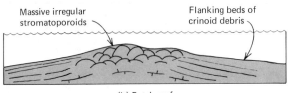

Massive irregular
stromatoporoids

Flanking beds of
crinoid debris

(b) Patch reef

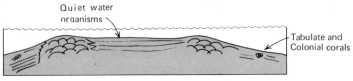

Quiet water
organisms

Tabulate and
Colonial corals

(c) Bank formed of coalesced patch reefs

FIG. 4-20 Types of middle Paleozoic carbonate buildups. (a) Lime mud mound. Bryozoans, algae, and sponges commonly provided the small amount of skeletal material. This type of mound frequently formed the base for larger, more complex mounds. (b) Patch reef. Stromatoporoids capped a mud mound. Crinoids growing on top and sides of a patch reef contributed large quantities of debris to the flanks of the buildup. (c) Patch reefs sometimes coalesced to form large banks. A variety of quiet-water organisms flourished in the lagoons between individual buildups on the bank. (Wilson, 1975)

quantities of gas and oil, developed in this way. Individual buildups coalesced into large barrier complexes and atolls at the margins of continental shelves and around major cratonic basins. Shallow portions of these large reef complexes were capped by stromatoporoids whereas the quiet waters of the interior lagoons supported a diverse fauna of more delicate organisms (Fig. 4-20).

Corals were severely reduced in the major extinction event at the end of the Devonian Period. Following this, carbonate buildups were again characterized by the lack of a rigid framework. In the Mississippian these buildups consisted of lime mud mounds that formed in moderately deep water between shallow carbonate shelves and adjoining basins. Despite the lack of framework, some mounds achieved a length of several kilometers, relief of more than 100 meters, and steep sides (30 to 50°).

Most Pennsylvanian and Permian reefs also seem to have started as accumulations of lime mud. Both rugose and tabulate corals were present, but only as minor constituents. Platelike growth forms of both red and green algae grew on the mounds in their early stages and trapped lime mud, aiding the mounds' upward growth. When the mounds encountered wave base, their tops became covered with binding and encrusting organisms, such as tubular foraminifera, algae, and stromatoporoids. Calcareous sponges were also common. Skeletal material eroded from the top of the buildups produced flank debris that typically accounted for a major portion of the reef's bulk. One of the most spectacular of all ancient reefs is the Permian Capitan reef of West Texas, which grew at the edge of a vast shelf that bordered a deep (600 meters) marine basin (Fig. 4-21).

FIG. 4-21 Permian of West Texas. (a) Middle Permian: well-developed basins rimmed by barrier reefs. Shallow marine limestones, evaporites, and red shales accumulated behind the reefs. (b) The Capitan Limestone reef front in the Guadalupe Mountains. Vertical relief is about 600 meters. (c) Cross section A–A' through the basin margin (King, 1948).

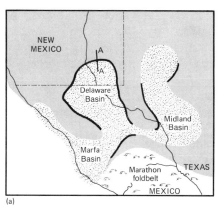

(a)　　　　　　　　　　　　　　　　　(b)

—— Reef tract　　　　　　　　　　☐ Platform limestone, anhydrite, and red shale

▓ Basin sandstones and limestones

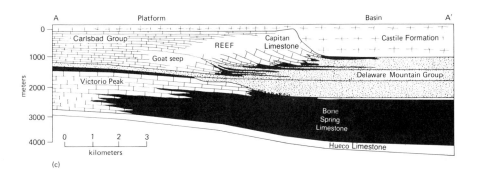

(c)

The Origin of Dolomite

Many Paleozoic carbonate deposits consist almost entirely of the mineral dolomite (calcium magnesium carbonate). Yet dolomite is not being deposited directly anywhere today. It is not secreted by any organism, modern or ancient, and it does not precipitate from seawater during evaporation to dryness. The failure to precipitate dolomite in the laboratory has confirmed the fact that it does not precipitate directly at surface temperatures and pressures, even from supersaturated solutions. For this reason, the origin of the vast deposits of dolomite in the geologic record has long been a mystery and is still not well understood. The explanation must lie in the *postdepositional* alteration of calcium carbonate minerals to dolomite and this process occurs in a variety of geological environments.

Today dolomitization occurs mainly in arid, marginal marine settings, such as the supratidal sabkha environments of the Persian Gulf and the Red Sea. (Sabkhas are extremely arid tidal flats flanking the Persian Gulf and the Red Sea on which salts are being deposited.) Strong winds coupled with extra-high tides occasionally flood the supratidal flats with seawater. As aragonite, gypsum, and anhydrite precipitate under extreme evaporative conditions, calcium is removed from the water and the ratio of magnesium to calcium increases. In the presence of the magnesium-rich brine carbonate sediments are altered to dolomite. Even if the associated evaporites are later altered or dissolved, relict evaporite fabrics in an ancient dolomite should provide a key to this kind of environment.

Under some circumstances the heavy brines formed in the evaporative tidal flat environment may also percolate down through older porous carbonate rocks below, altering them to dolomite. The Silurian pinnacle reefs of the Michigan Basin appear to have undergone this type of reflux dolomitization as tidal flats developed over the tops of the old reefs and the adjoining basinal sediments during late stages of the evolution of the basin. The selective dolomitization of reefs in the geologic record probably results mainly from their strength and resistance to compaction, which preserves their high initial porosity, and partly from red algae, which secrete calcite skeletons that are rich in magnesium and that have long been a major component of reefs.

At present, dolomite is also forming in regions that are alternately humid and dry. In the Coorong area of South Australia dolomite forms in ephemeral coastal lakes that fill during the rainy season due to groundwater flow from carbonate aquifers. The lakes evaporate during the dry season when the water table sinks below the surface on the lakes. Here the source of the magnesium required for dolomitization is apparently the carbonates in the aquifers and volcanics in the recharge area rather than seawater. The percentage of dolomite is highest in lakes that are farthest inland and above the zone of seawater-freshwater mixing. There is no evidence of any evaporative minerals associated with these dolomitic sediments; apparently any evaporites formed in the lakes are flushed from the sediment during the wet season and carried seaward in the groundwater.

In the Coorong environment very little dolomite forms in the zones where seawater is in contact with meteoric water, but in other regions dolomitization of carbonates within this zone has probably been an important process. In this model seawater is the source of the magnesium and the mixing of seawater with fresh groundwaters creates a fluid in which dolomite is supersaturated and calcite is undersaturated. The result is dolomitization of the carbonates. The setting for this type of dolomitization is a geologically common one in which fresh groundwater moves seaward and mixes with seawater within relatively porous carbonates in coastal areas (Fig. 4-22). Relative lowering of sea level would expose a greater area of the carbonate shelf to meteoric recharge and move the zone of dolomitization seaward. Rising sea level would inundate shelf carbonates, covering the recharge area with seawater and driving the zone landward. So at a particular location vertical changes in the degree of dolomitization may occur because of shifts in the extent of fresh groundwater as a result of transgression and regression. The Michigan Basin pinnacle reefs appear to have undergone this type of dolomitization in an early stage, before the reflux dolomitization just described. A fall of sea level exposed the tops of the reefs and permitted the development of karst topography and solution porosity on the reef tops. Within the reef, dolomitization took place in the mixing zone between meteoric and seawater. Once the basin was nearly filled with sediment, the refluxing brines from the supratidal flats completed the dolomitization of the reefs during a subsequent rise in sea level.

Reef limestones may undergo further dolomitization after burial by providing a pathway for magnesium-rich waters expelled by adjacent shales during compaction. Magnesium is released to interstitial brines by certain clay minerals during diagenesis. (*Diagenesis* refers to any post-depositional chemical reactions between grains of sediment and the solutions around them.) The

FIG. 4-22 Diagrammatic representation of a modern (Pleistocene) example of dolomitization taking place in the mixing zone between meteoric and marine waters. (Land, Blackwell Scientific Publications Limited, 1973)

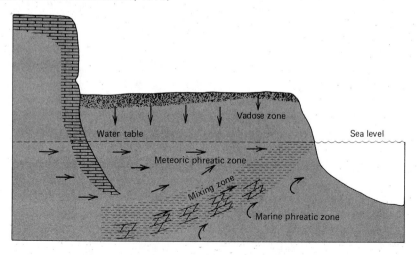

magnesium-rich brines flow from the compacting shales through adjacent porous and permeable reefs, which are progressively dolomitized in the diagenetic environment (Fig. 4-23).

Evaporites

At times during the Paleozoic widespread evaporite sediments—chiefly calcium sulfate (the minerals gypsum and anhydrite) and sodium chloride (the mineral halite)—were deposited in the shallow epicontinental seas. Evaporite deposits are unique in that they require no physical transport of sedimentary particles from afar; nor do they need to be produced by living organisms at the site of deposition. Instead they precipitate directly, and sometimes rapidly,

FIG. 4-23 The highly permeable and relatively incompressible reef provides a convenient pathway for magnesium-rich waters squeezed from adjacent strata as they undergo compaction. In time they cause the reef to become dolomitized.

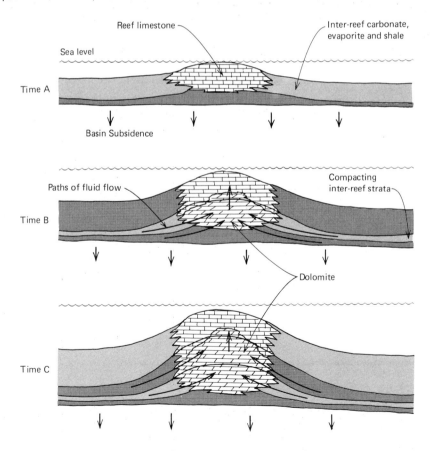

from normal seawater as it evaporates nearly to dryness. Gypsum precipitation begins after 80 percent of the seawater has evaporated and halite precipitation after 90 percent has evaporated. Such intense evaporation requires an area with an arid climate and impeded circulation with the open sea. These conditions occurred repeatedly throughout the geologic past and produced large deposits of salts.

The thick Silurian deposits of the Michigan Basin, for example, include 750 meters of evaporite strata (mostly halite and anhydrite). If this much salt were to form by the simple evaporation of a standing body of seawater of normal salinity (35 parts per thousand), the water would need to be about 1000 kilometers deep! This, of course, is unreasonable and other evidence suggests that the depth of the Silurian sea in the Michigan Basin was never more than a few hundred meters, if that. Indeed, considerable controversy surrounds the interpretation of the depositional environments of the reefs, evaporites, and associated bedded carbonates of the Michigan Basin. Substantial discoveries of petroleum and natural gas in these rocks have spurred interest in this controversy.

During the Silurian a large complex of reefs, bedded carbonates, and evaporites formed in what is now the Great Lakes region. Barrier reef deposits up to 200 meters thick form a rim around the Michigan Basin in which thick sequences of evaporites and bedded carbonates were deposited (Fig. 4-24). Within the barrier reef belt lies a belt of pinnacle reefs, 6 to 10 kilometers wide. The question is whether the evaporites in the middle of the basin formed in shallow or comparatively deep water. These who favor the deep-water model consider the Michigan Basin a classical example of a barred basin in which the evaporites formed in water that may have been up to a few hundred meters deep behind the encircling reef. In this view, high salinities were achieved by intense evaporation; the high-salinity water did not mix readily with seawater of normal salinity in the open sea beyond because mixing was retarded by the encircling barrier reef. Saturation was reached, evaporation continued, and the salts precipitated. Several inlets have been identified in the reef and these openings could have provided a restricted supply of new seawater (and salts) to the evaporite basin. Evidence for the continuous growth of basin margin reefs during the time of evaporite deposition strongly favors this model. Yet the importance of the barrier to evaporite formation seems questionable because several thick layers of salt formed after the barrier reef was completely buried and no longer had any influence on sedimentation.

The shallow-water interpretation of the Michigan Basin evaporites depends on large-scale drops in sea level within the basin in order to produce sabkha-type evaporites. In the sabkha model evaporites precipitate beneath supratidal salt flats that are periodically inundated with seawater during storms and unusually high tides. In the modern-day sabkhas of the Persian Gulf aragonite and gypsum precipitate beneath algal mats that cover the seaward parts of the sabkha. In the upper reaches of the flats gypsum is altered

FIG. 4-24 Silurian paleogeography of the Great Lakes region. Reefs grew around the margins of the Michigan Basin. A thick sequence of evaporites and limestones was deposited within the basin. Later the entire area was covered with another thick layer of evaporites. (Mesolella, 1978)

diagenetically to anhydrite within the capillary zone. The anhydrite grows in the form of nodules that displace the surrounding sediment and produce a distinctive evaporite fabric known as "chicken wire anhydrite." The geometry and sequence of the evaporite units in the Michigan Basin suggest that periodic inundations of the sea were separated by periods of intense evaporative drawdown. Proponents of the shallow-water model point to evidence for subaerial exposure of pinnacle reef tops and to facies changes in the evaporites (from thick, basin-center salt to thin, evenly bedded anhydrites, to thin, nodular anhydrites at the basin margin) to support their conclusion that the evaporites formed in shallow water after all or most of the reef growth had ended due to fluctuations in the sea level.

Glacial Deposits

During the late Paleozoic huge continental glaciers formed on the Gondwana continent. In thickness and extent these must have rivaled the huge ice sheets that formed during the great Pleistocene glacial ages of the last 2 million years. The Paleozoic glacial deposits contain most of the characteristic features common in Pleistocene glacial deposits.

1. Boulder-laden shales, inferred to be tillites, contain striated and faceted pebbles and they rest on grooved and polished bedrock below (Fig. 4-25).
2. Regionally several successive tillites are separated by marine deposits, which suggests alternating glacial and interglacial episodes.
3. Banded shales, also interbedded with the tillites locally, may represent varved clays that were deposited in proglacial lakes.
4. Linear bodies of sand locally interbedded with the tillites may represent buried eskers.

Studies of the orientation of glacial striations and *roches moutonnées* indicate that a single large ice sheet originated in two centers, one in southwestern Africa and another in eastern Antarctica (Fig. 4-26). Both centers are near the Permo-Carboniferous South Pole as determined from rock paleomagnetism. The wide distribution of the glacial deposits and the radial pattern of the ice movement indicate continental rather than alpine glaciation. Glaciers apparently extended to within 30° of the equator at the time. More recently, Pleistocene ice sheets reached to within 40° of the equator. Thus the extent of the late Paleozoic and the Pleistocene glaciations is comparable.

FIG. 4-25 Permian glacial deposits in southern Australia. Grooves and striations on the rock floor beneath the tillite confirm its glacial origin. (Courtesy of W. C. Bradley)

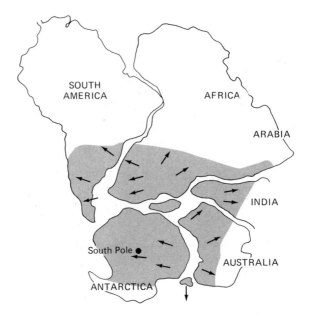

FIG. 4-26 Inferred distribution of Pennsylvanian and Permian ice sheets. Arrows show direction of ice flow determined from glacial pavements.

Late Paleozoic Cyclic Sedimentation

Strata in the cratonic interior of North America display a striking pattern of repetitive cycles throughout the Pennsylvanian and into the Lower Permian. Similar cyclic sedimentation also occurs in late Paleozoic strata on other continents. In North America the emergent Appalachian region shed sediments westward onto a vast low coastal plain. This plain was periodically invaded by the shallow sea that occupied the interior of the continent. The periodic invasions and retreats of the sea are the immediate cause of the sedimentary cycles, which are called *cyclothems*.

In Kansas the cyclothems consist predominantly of rocks of marine origin. In Illinois they form about equal portions of marine and nonmarine strata, including widespread and important coal beds. Near the eastern source area, from Pennsylvania southward to Tennessee, the Pennsylvanian and Permian cyclothems are composed chiefly of nonmarine sandstones alternating with coal beds.

At least 50 late Paleozoic cyclothems are known, many of which can be traced widely. A typical cycle in the Illinois region (Fig. 4-27) begins with cross-stratified, poorly sorted sandstone that typically rests on a channeled surface in the underlying beds. These basal sandstones are interpreted as river and delta deposits. The coals in the middle of the cyclothems apparently

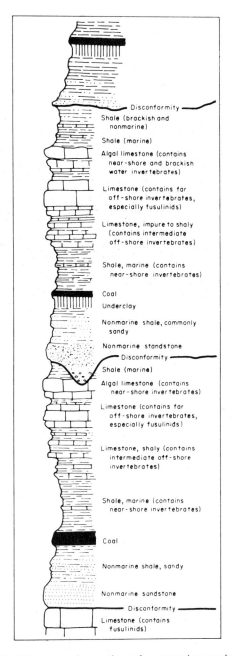

FIG. 4-27 Diagrammatic section of successive cyclothems. (Matthews, 1974 after Moore, 1964)

formed in vast coastal swamps containing junglelike vegetation similar perhaps to the Florida Everglades that today cover 25,000 square kilometers. The late Paleozoic coal swamps must have been much larger, however, for some coal beds can be traced over vastly greater areas. The ancient swamps were dominated by large scaly-barked trees (lycopsids), but they contained many smaller kinds of plants as well. The trees and shrubs that died and fell to the swamp floor were protected from oxidation by stagnant waters and rapid burial. In time the accumulated organic material was compacted by the weight of subsequent sediments to form peat and, later, coal.

The upper part of an Illinois cyclothem generally contains fossiliferous marine shales and limestones. They indicate that the sea flooded the low coastal swamps; the marine upper phase of each cyclic set represents the peak of transgression. While the cyclothems were being produced, the central part of the craton must have been exceedingly flat and very near sea level so that a small change in sea level could produce a very large movement of the coast line.

The marine beds at the top of each cycle are, in turn, overlain by basal fluvial sandstones that mark the beginning of the next successive cyclothem. The approximately 50 cyclothems were thus produced by as many transgressions and regressions of the sea over a period of about 50 million years. It represents a period of unusually rapid transgressions and regressions. This unusual phenomenon was the key to the vast quantities of coal produced during this time, coal that provided the primary energy source for the industrialization of the United States and Europe.

Some transgressions and regressions that produced the cyclothems may have been generated locally within the depositional basin—regressions when deltas grew temporarily seaward and transgressions whenever points of river discharge shifted to another coastal area. The regularity and apparent synchroneity of most of the cycles throughout the depositional basin, however, suggest that they resulted from numerous eustatic sea level changes. The reason for these apparent sea level changes is unknown. When these unusual deposits were forming, however, extremely widespread glaciation was occurring in the southern hemisphere and possibly the two events are connected. During the ice ages in the most recent 2 million years of geologic history continental ice sheets did not simply form once, stabilize themselves on the continents for the duration of the Pleistocene Epoch, and then disappear. Instead the ice sheets waxed and waned several times, producing sea level fluctuations of about 150 meters. A similar waxing and waning of the late Paleozoic ice sheets may have caused the eustatic sea level changes that, in turn, produced the cyclothems. On a coastal plain of low relief, having a slope of 0.25 meter/kilometer, a 200-meter sea level change would cause the sea to transgress and regress over a distance of 800 kilometers (500 miles), more if isostatic adjustment is taken into account.

Clues to Paleozoic Postdepositional Environments

Postdepositional environments are the sites of physical and chemical changes that occur within sediments as they undergo processes of compaction and diagenesis. Some of these processes occur very early—after the sediments are buried only a few centimeters. Others occur much later as heat and pressure increase with deep burial. Proper interpretation of these diagenetic environments provides information not only on chemical conditions to which sediments have been subjected but also on former quantities of overburden that may have been long since removed.

As rocks are buried to great depth, the increasing heat encountered alters any finely divided organic constituents that they may contain. In particular, complex organic molecules tend to break down to simpler molecules that are important constituents of petroleum and natural gas. Thus rocks that are especially rich in organic material are potential source rocks for hydrocarbons. The breakdown from complex to simpler organic molecules requires (1) heat and (2) time. Pressure plays little or no part. The greater the heat and the longer the rocks are subjected to it, the greater are the changes. Consequently, the extent of these changes is referred to as a rock's *thermal maturity*. Young rocks that have not been heated significantly are immature and generally incapable of yielding any hydrocarbons. Organic molecules within the sedimentary rocks change to compounds of petroleum and natural gas with increasing heat and time. With additional heat, the upper threshold for oil generation is soon reached; beyond this threshold, through a substantial range of temperature, the rocks produce only gas. Finally, an upper threshold for gas generation is also reached. So an understanding of the thermal history of a body of sedimentary rocks can be used to evaluate its potential for producing hydrocarbons well in advance of actual drilling.

In addition to the actual petroleum or natural gas that a potential rock may contain, other minor ingredients in the rock may provide a clue to its thermal history. In some cases, relatively large fragments of shiny black organic material (vitrain) that can actually be observed under the microscope can be used as an indicator of thermal maturity. Some sedimentary rocks contain either palynomorphs (spore and pollen grains) or conodonts (tiny toothlike objects 0.5 to 1.0 millimeter long) that exhibit systematic color changes corresponding to the thermal maturation of the rock in which they occur. Spore and pollen grains are themselves composed of carbon compounds and are rather easily oxidized. Thus they tend to be poorly preserved in rocks as old as the Paleozoic. Conodonts, on the other hand, consist chiefly of the very resistant mineral apatite (calcium phosphate) with a substantial content of organic material. Conodonts range only from the Cambrian to the Triassic and so they provide convenient indicators for Paleozoic and Triassic rocks and nicely complement the palynomorphs that find their best use in rocks of later Mesozoic and Cenozoic age.

Totally unaltered conodonts are pale yellow and with increasing heat—starting at about 50°C—they gradually turn brown and then black. As heat increases, above about 350°C, carbon is progressively lost; consequently, conodonts change to an opaque gray color and then finally become clear. As temperatures approach 600°C, well into the range of medium-grade metamorphism, conodonts are destroyed. Each perceptible color change is numbered and conodont alteration indices, as they are called, range from 1 for light brown specimens to 5 for the clear conodonts that are the most altered of all. A conodont alteration index (CAI) of 2 represents the upper limit for preservation of compounds of petroleum. A CAI of 4 represents the approximate upper limit for natural gas. Rocks bearing conodonts with a CAI greater than 4 are generally devoid of hydrocarbons. Figure 4-28 shows an example of the use of conodonts as indicators of thermal maturity in the Ordovician strata of New York, northern Pennsylvania, and northeastern Ohio. The conodont alteration index increases from 1.5 on the west to 5 on the east. This gradient reflects depth of burial and hence the temperatures to which the Ordovician rocks of this region have been subjected. Isopach maps of the overlying Silurian and Devonian strata show a pronounced thickening to the east and verify this interpretation. The CAI of 2 and 4, which are upper limits for production of oil and natural gas, respectively, correspond to the eastern limits of Ordovician oil fields and gas fields in the region.

FIG. 4-28 Conodont color alteration index (CAI) isograds in Ordovician carbonate rocks of New York State show thermal maturity increasing eastward. (After Harris, Harris, and Epstein, 1978)

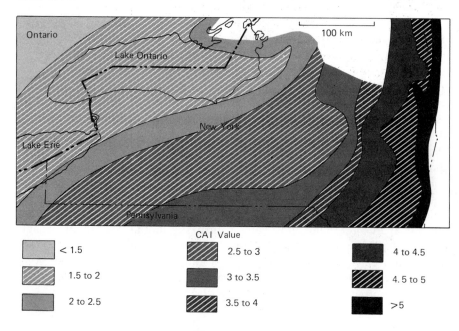

CAI Value		
< 1.5	2.5 to 3	4 to 4.5
1.5 to 2	3 to 3.5	4.5 to 5
2 to 2.5	3.5 to 4	>5

PALEOZOIC LIFE

The Diversification of Life

The numbers and types of living organisms present on the earth expanded enormously during the 350 million years of Paleozoic time. After some 4 billion years, during which the earth's surface was almost devoid of animal life, all major groups of shelled invertebrates appeared in abundance in the rocks of the Cambrian and Ordovician systems. No new phyla and only a few classes of invertebrate animals have evolved since the Ordovician. At the beginning of the Paleozoic sea-dwelling algal plants and invertebrate animals dominated the living world and the land surface was devoid of life. Then in the Silurian Period water-dwelling algae gave rise to simple mosslike and fernlike plants that, in turn, evolved into the earliest trees and shrubs. Apparently they covered much of the land surface by mid-Paleozoic time (Fig. 4-29). Plants — and later animals — could have colonized the land only if the atmospheric concentration of ozone was sufficient to shield the land surface from the lethal

FIG. 4-29 The Paleozoic origin and expansion of land plants and animals, all of which evolved from water-dwelling ancestors.

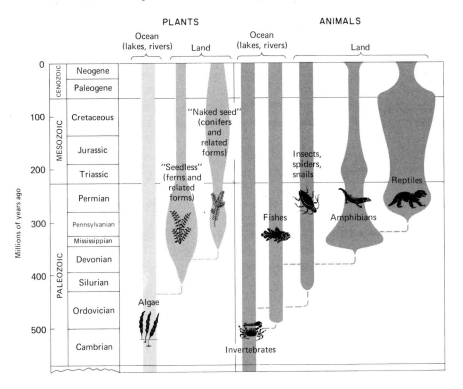

ultraviolet radiation. An effective ozone shield requires that the amount of oxygen in the atmosphere be at least 1 percent of its present concentration.

Accompanying the spread of Paleozoic land plants was the nearly simultaneous rise of land animals from water-dwelling ancestors. By far the most successful land animals, measured by numbers of individuals and species, are the insects and arachnids (spiders and related forms) of the phylum Arthropoda. Both groups evolved in mid-Paleozoic time along with the first land plants and they have subsequently increased to their present abundance (Fig. 4-29). At about the same time that these groups evolved, certain groups of water-dwelling snails (phylum Mollusca) also developed the ability to breathe air and left the water to feed on land plants; they are still common today.

The final phylum that includes land-dwelling animals, the phylum Chordata, is the most interesting to us because it contains all land-dwelling vertebrate animals, among them our own species *Homo sapiens*. The earliest fossil vertebrates are primitive fish found in Upper Cambrian deposits (Fig. 4-29). These fish, the ostracoderms, were heavily armored and fed by grubbing in the mud on the bottom of the ocean. By Devonian time a great variety of fish lived in both freshwater and the sea. Some were lightly built, fast swimmers. The bones in their fins were thin and arranged in a diverging pattern. The fins were used mostly for steering while their tails provided propulsion. They were the ancestors of most modern fish.

Another group of freshwater fish had heavier bodies and sturdy, powerful fins (Fig. 4-30). The bones in their fins were thick and were aligned in an arrangement reminiscent of the bones in a hand or foot. Their upper jaws were hinged, which made biting prey in shallow water easier because the jaw could open upward out of the water rather than downward into the mud. Modern crocodiles, living in similar shallow swamp and stream environments, have this same adaptation. These *lobe-finned* fish lived in ephemeral ponds and streams; when the water was low, the sturdy bone structure of their fins helped

FIG. 4-30 Devonian lobe-finned fish were the ancestors of the first amphibians. (Courtesy of the American Museum of Natural History)

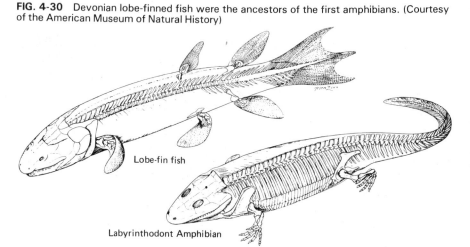

Lobe-fin fish

Labyrinthodont Amphibian

to support them as they crawled to another pond. These fish were the ancestors of the amphibians (Fig. 4-30), the first vertebrates to inhabit the land.

The plants and animals that first invaded the land adapted to a new and very different environment. They needed new forms of support and locomotion, protection from drying out, and new means of obtaining gases required for photosynthesis and respiration. The density of air is much less than that of water and thus air does little to help support an organism living on the land. Many organisms live in the upper levels of the sea, but how many organisms can you think of which actually live permanently suspended or "swimming" in the air? Plants developed woody stems and the fins of fish evolved to legs to support their land-dwelling descendents, the amphibians. Most terrestrial animals have legs because the legs raise most of the body surface up off the ground, thereby greatly reducing the amount of friction and enhancing mobility.

Animals and plants leaving the aquatic environment also faced the problem of drying out. The development of scales and a special egg with a tough semipermeable covering that kept water in while oxygen and carbon dioxide passed through were major advances that made the reptiles (Fig. 4-31) the first completely terrestrial vertebrates. The first reptiles arose in the Pennsylvanian and by the close of the Permian a host of large, herbivorous and carnivorous reptiles roamed the land. Plants developed waxy coatings and special cells that permitted the exchange of gases without water loss. And now that these plants were no longer suspended in water, surrounded by dissolved nutrients, roots evolved, both for support and to extract nutrients and water from the soil.

FIG. 4-31 A carnivorous Permian reptile. (Robert Bakker Reconstruction)

The first invertebrates to invade the land were scorpionlike arthropods. Their tough, impervious external covering and jointed legs were already well adapted to protect them from drying out and to provide support and easy locomotion in a terrestrial environment. Even today the vast majority of terrestrial invertebrates are arthropods.

Freshwater lakes and ponds tend to be small and poorly oxygenated compared to most oceans. Air is a much better source of free oxygen for respiration than water (21 percent in the air compared to a maximum of about 0.6 percent for very cold water), but gills do not function well in air. They collapse and stick together as they dry out, reducing the surface area available for gaseous exchange. Poor oxygenation of ponds and streams may have conferred a considerable advantage on fish that were able to tap the oxygen resource of the air and many early freshwater fish probably had lungs. Lungs were only the first step in the evolution of a whole complex of adaptations for air breathing, which also involved dramatic changes in respiratory physiology and the circulatory system.

Permian Extinctions

The close of the Paleozoic Era was marked by extraordinarily severe and widespread extinctions of preexisting life. About 30 percent of the families of fossil animals and plants found in Lower Permian rocks became extinct by the close of Permian time and are unknown in younger rocks (Fig. 4-32). Marine invertebrate animals were particularly hard-hit. Trilobites became extinct, as did many previously abundant groups of corals, bryozoans, brachiopods, and crinoids. In broad terms, deposit-feeding invertebrates were less affected than suspension feeders; and invertebrates that were tolerant of a relatively wide range of salinities were less affected than invertebrates that required normal marine salinities.

The extinction at the close of the Permian was one of several that occurred during the Phanerozoic. The reason for such massive extinctions remains elusive and is the source of much speculation. The answer must lie in some environmental change on a global scale. But the same kind of environmental change was not necessarily responsible for all the major extinctions in the history of the earth. Moreover, the cause of any one of them may not have been a singular event but the unusual coincidence of several unfavorable environmental factors. No one explanation of massive extinctions has received widespread acceptance. All are open to challenge because there is no direct evidence for the exact cause. We can observe which groups of organisms with hard parts were most or least affected, but we cannot perform autopsies on the fossils. If we could, we might be able to say "They all starved to death," or "They died of asphyxiation due to low levels of oxygen," or "They were poisoned by that toxic substance." Even then we would lack the complete picture, for we would not know why they starved or what the source of the poison was. Some very plausible causes might not leave any direct evidence, for extinctions

FIG. 4-32 Phanerozoic expansions and extinctions of life. The curves show net expansions (light color) or extinctions (dark color) of animal and plant families during each Phanerozoic period.

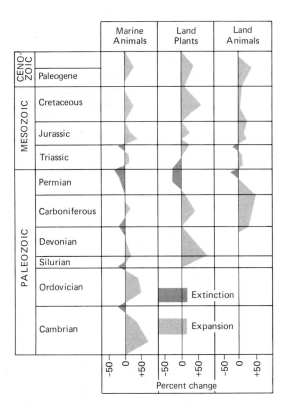

of most species probably result from failure to reproduce rather than massive death.

The preferential survival of detritus feeders over suspension feeders suggests a drastically reduced plankton population. Some geologists have suggested that by the end of the Permian the supercontinent had been eroded to a flat-lying land that contributed little in the way of sediment or nutrients to the sea and that global nutrient depletion led to the decimation of the plankton. Such an earth is difficult to imagine. Renewed tectonic activity took place on the western margin of North America in the Late Permian and major uplifts accompanied the final suturing of the continents into Pangaea. These uplifts must have provided major sources of sediment and relief for drainage to the sea.

Indeed, the assembly of the supercontinent may have actually been responsible, at least in part, for the Permian extinctions. Individual continents that are separated by wide deep oceans have their own individual terrestrial and shelf fauna and flora. As the elements of the supercontinent were assembled, these separate biotas from each element were forced into competition. The number of sets of niches was drastically reduced and many species must have become extinct from competitive exclusion. At the same time, the climatic effects of increasingly great thermal fluctuations on the supercontinent would have created less stable environments, which could support fewer, less specialized species.

The salinity of the world ocean may have become unusually low by the end of the Permian. Large quantities of salt were deposited in evaporitic environments during the Permian. We do not know exactly how much because significant quantities have probably dissolved since Permian time and large quantities of Permian salt may still be unknown or improperly dated. Known Permian salt deposits contain an amount of salt equal to at least 10 percent of all the salt dissolved in the present ocean. The removal of this amount of salt from the ocean today would lower the average salinity to about 31.5 o/oo, about half the drop necessary for large-scale deleterious effects on stenohaline marine invertebrates. If significantly larger volumes of salt were actually deposited than remain today, of if refluxing brines carried large quantities of salt to the deeper parts of the ocean, the salinity of surface waters may have fallen low enough at the end of the Permian to cause the extinction of a large number of stenohaline invertebrates.

The massive deposition of evaporites might have had a significant effect on the atmosphere as well as the ocean. The amount of oxygen tied up in Per-

FIG. 4-33 A comparison of extinction rates at the family level with oxygen consumption rates for the closest modern relatives shows a strong correlation between extinction and high levels of oxygen consumption.

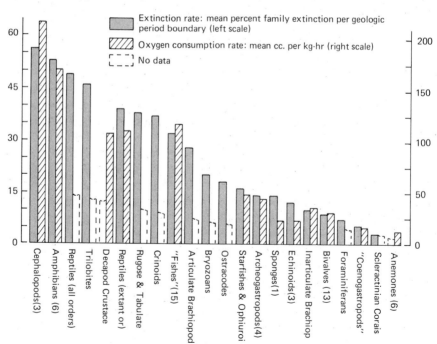

mian sulfates (SO_4^2) is estimated to be equal to 90 percent of the oxygen present in the atmosphere today. If low oxygen levels were responsible for Permian extinctions, we would expect those animal groups with the highest oxygen requirements to be most affected. A comparison of extinction frequency of major groups with what we know of their oxygen requirements does appear to show a significant correlation (Fig. 4-33). It might also explain why the warm-blooded (endothermic) mammallike reptiles were the major terrestrial organisms to be affected by the Permian extinction, for endothermy requires a high level of oxygen consumption. Yet the correlation between extinction frequency of major animal groups and their oxygen requirements rests on skimpy data and may be partly coincidental. Indeed, measurements of oxygen consumption used for comparison among animal groups must, of necessity, be made on the descendents of groups that *survived* the extinction events. None of the ones that died out is left to test!

the mesozoic earth

The Mesozoic Era of earth history lasted about 160 million years, less than half as long as the Paleozoic (Fig. 5-1). In spite of its relatively short duration, the Mesozoic was a time of extraordinary changes in the earth's geography, life, and environments. As it began, the forerunners of today's major landmasses lay joined together in one great continent, Pangaea. This huge continent apparently stood high with respect to the sea, for most of the early Mesozoic sediments preserved today, like those of the latest Paleozoic, were deposited in terrestrial environments.

MESOZOIC GLOBAL TECTONICS

The Breakup of Pangaea

Our modern world had its beginning late in the Triassic when the continent of Pangaea began to break apart. Evidence for the timing of the Mesozoic breakup of Pangaea and for the subsequent migrations of the continents is provided primarily by rock magnetism. Rock magnetism is caused by the alignment of numerous tiny grains of magnetic minerals in igneous or sedimentary rocks and the alignment itself was produced by the earth's magnetic field at the time the rock formed. The rock record of ancient earth magnetism, called *paleomagnetism*, enables us to determine the position of the North and South poles for successive times in the geologic past. This record indicates that all the continents continually changed position relative to the poles through geologic time. The change in polar position plotted on the globe from the point of view of a given continent is that continent's *polar wandering path.*

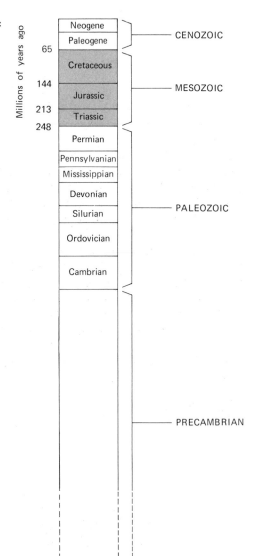

FIG. 5-1 Position of the Mesozoic Era in the geologic time scale.

The earth has always had a single south and north magnetic pole so that if two continents have maintained the same position with respect to each other, their polar wandering paths should coincide. Yet continents today do not share the same polar wandering paths, which can only mean that they have moved relative to one another. Figure 5-2 shows that while the American and European north pole positions coincide now, they diverged progressively in the geologic past. The indicated north poles for a given time can be brought into concordance by pushing the continents back together progressively until, in the

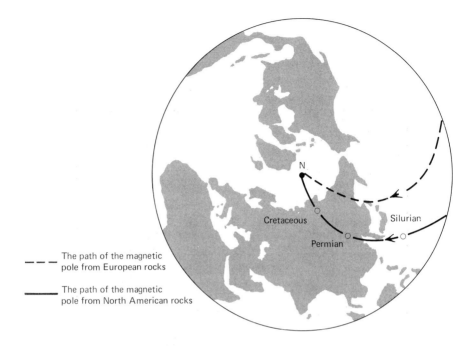

FIG. 5-2 Polar wandering curves for Europe and North America.

Triassic Period, they join and the Atlantic Ocean is closed. Polar wandering paths for other continents similarly demonstrate the timing of their separation from Pangaea.

Additional evidence for the timing of the breakup of Pangaea comes from the stratigraphy of the deep ocean basins. As continents move apart, ocean basins are created, the new crust forming continually at oceanic ridges. So we would expect to find the youngest oceanic crust at the oceanic ridges and the oldest crust in those parts of the ocean basins that are farthest from the ridges. Deep drilling in the oceans has verified that progressively nearer the oceanic ridges, the oldest sediments, immediately over the basaltic crust, are progressively younger. This is excellent evidence that the ocean floor itself decreases in age toward the ridges. The oldest North Atlantic sea floor sediments, for example, are in the extreme western and extreme eastern parts of the Atlantic and are Jurassic in age (Fig. 5-3).

The Displacement of the Continents

Figure 5-4 shows the changes in continental configurations during the Mesozoic, based on the kinds of evidence just enumerated. When the Triassic Period began, the amount of land on the earth's surface was nearly equally distributed between the northern and southern hemispheres. Today, after prolonged separation of the fragments of Pangaea, two-thirds of all land lies

Pleistocene to Holocene	Eocene
Pliocene	Paleocene
Miocene	Cretaceous
Oligocene	Jurassic

FIG. 5-3 Age of the North Atlantic Ocean floor decreases toward the Mid-Atlantic ridge. (Pitman, Larson, and Herron, 1974)

north of the equator. The northward displacement removed the present southern continents (excepting Antarctica) from the high-latitude positions that had promoted their widespread glaciation in the late Paleozoic. The present northern continents simultaneously moved north, away from the equator to achieve their present positions.

Triassic basalt flows and fault basins of the eastern coast of North America record an important episode of rifting between the northern and southern portions of Pangaea 200 million years ago. The rifting initially occurred along a nearly east–west trend not far north of the Triassic equator (Fig. 5-5). Late Triassic and Early Jurassic evaporites formed in the barely open North Atlantic, which at one point must have looked much like the Red Sea today. During the Jurassic Period similar evaporites formed in what is now the Gulf of Mexico. Still later, Early Cretaceous rifting of the South Atlantic created a comparable evaporite basin and produced salt deposits that today underlie the

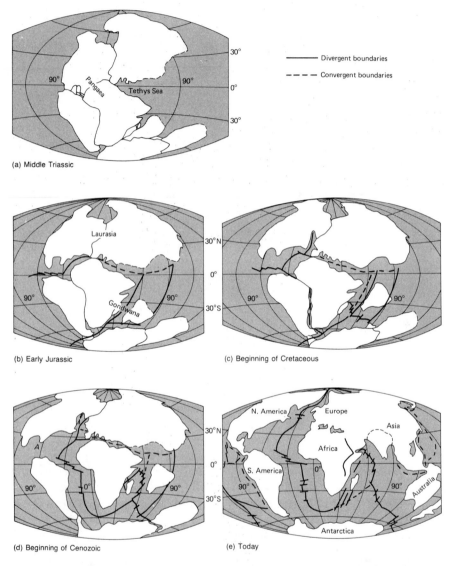

(a) Middle Triassic

(b) Early Jurassic

(c) Beginning of Cretaceous

(d) Beginning of Cenozoic

(e) Today

——— Divergent boundaries

– – – Convergent boundaries

FIG. 5-4 The Mesozoic breakup of Pangaea. (a) Earliest Triassic: the single continent of Pangaea. (b) Early Jurassic: separation of Laurasia from Gondwana, the eastern part of which has begun to break up. (c) Early Cretaceous: North Atlantic and Indian oceans widen. Breakup of eastern Gondwana continues. (d) End of Cretaceous: South America has separated from Africa. (e) The modern continents: Europe and North American are completely separated, Australia has moved north away from Antarctica, and India has collided with Asia. (Dietz and Holden, 1970)

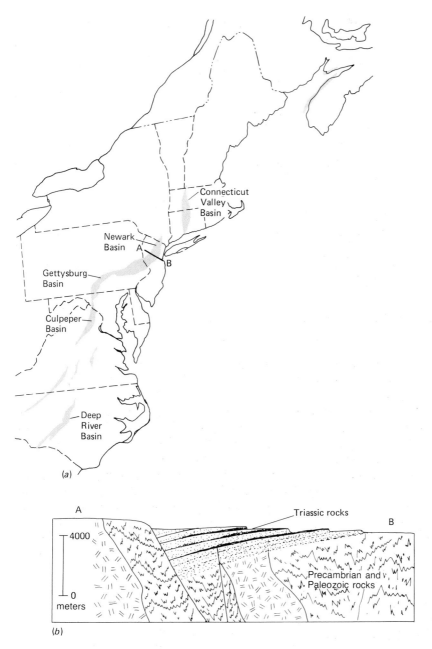

FIG. 5-5 The Triassic block-fault basins along the Atlantic coast of North America developed from tensional forces during the opening of the North Atlantic Ocean. They contain thick sequences of alluvial sedimentary rocks and basalt flows. The Newark Basin is shown as an example.

125

South American and African shelf areas indicated in Fig. 5-6. In all three regions the salt has since been locally squeezed upward through a thick layer of younger strata to form salt *diapirs*. These diapirs or salt domes created numerous oil traps at depth, some of which are now being exploited.

In the Late Triassic far to the south another major rift developed where South America and Africa began to separate as a single unit from the remainder of Gondwana [Fig. 5-4(b)]. Soon afterward the Indian Ocean began to form when India broke away from Antarctica and rapidly drifted northward. During the Jurassic the North Atlantic continued to widen while the Tethys Sea became progressively narrower as Africa and Eurasia rotated toward each other.

South America and Africa also began to separate in the Early Cretaceous. Rifting began in the south and progressed northward with time. Complete separation of South America and Africa finally took place in the Late Cretaceous. In the Late Cretaceous the Labrador Sea opened between Canada and Greenland. By the end of the Cretaceous the South Atlantic had widened into a major ocean 3000 kilometers wide and Greenland was the only connection between North America and Eurasia (Fig. 5-4).

FIG. 5-6   Shaded areas contain abundant salt diapirs indicating the extent of massive salt deposition, by the oceanic sources indicated, during rifting in the Jurassic and Early Cretaceous. (Burke, 1975)

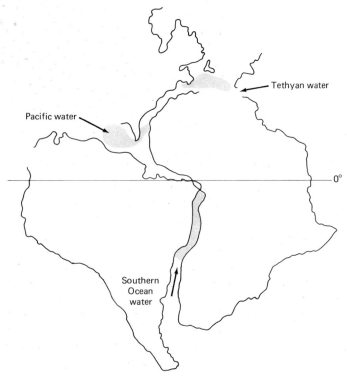

Cordilleran Orogenies

In southern and central California the Triassic continental margin truncates the southwest-trending belt of Paleozoic miogeoclinal rocks as well as the region of accreted Paleozoic terranes that lies immediately to the west (Fig. 5-7). It signals an episode of tectonic activity that began early in the Triassic with the removal of a portion of the old Paleozoic continental margin either by rifting or, more likely, transform faulting. Shortly afterward a volcanic arc formed along the western margin of North America and it persisted throughout most of the Mesozoic Era. In the Triassic the arc apparently consisted largely of volcanic islands. East of the arc in what is now Idaho and northern Nevada an uplift called the Mesocordilleran Highland began to grow. This renewal of tectonic activity is known as the *Nevadan Orogeny*.

In the Jurassic the broadening of the Mesocordilleran Highland created a

FIG. 5-7 In California Paleozoic and Precambrian depositional trends are truncated by Triassic oceanic rock assemblages. (Dickinson, 1979)

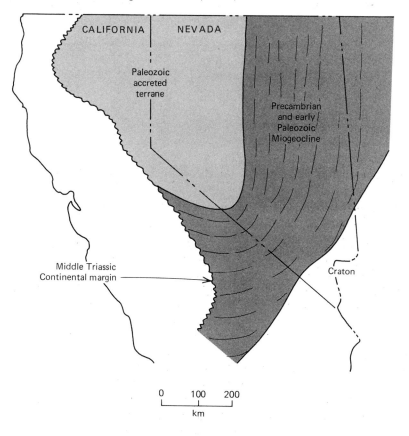

mountainous belt along the entire western margin of North America (Fig. 5-8). Volcanism was widespread along this belt and the preexisting volcanic island arc may have been largely replaced by a mainland volcanic chain much like the volcanoes in the present-day Cascades or the modern Andes of South America. Deformation was intense along the western margin of the continent as oceanic crust was subducted beneath the volcanic highland. During the latest Jurassic and the ensuing Cretaceous Period huge batholiths were intruded beneath much of the highland (Fig. 5-9). Some of these batholiths form the spectacular scenery of the Sierra Nevada mountain range. (Fig. 5-10).

The Mesocordilleran Highland continued to grow in the Early Cretaceous. Along its eastern margin thrust faults with several miles of eastward displacement formed during a major compressional episode called the *Sevier Orogeny*. Afterward huge volumes of clastic sediments were shed into adjacent oceans both to the east and to the west (Fig. 5-11). East of the highland, in the western interior of the United States, approximately 4 million cubic kilometers of Cretaceous sedimentary rocks accumulated. This accumulation, on the east side alone, represents the erosion of 8 kilometers of material from the entire area of the highland. In the latest Cretaceous and early Cenozoic widespread vertical faulting east of the Sevier Orogenic belt produced large uplifts from Canada to New Mexico. This episode of dominantly vertical movements is known as the *Laramide Orogeny*; during this time the major block uplifts that became today's Rocky Mountains (Fig. 5-12) were formed.

On the west, between the volcanic arc that was part of the Mesocordilleran Highland and the oceanic trench that lay approximately at the site of the

FIG. 5-8 Paleogeography of the Late Jurassic in North America.

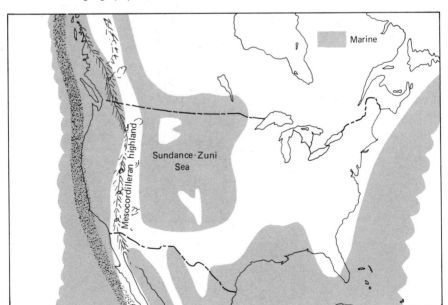

FIG. 5-9 Jurassic and Cretaceous grantic batholiths in western North America.

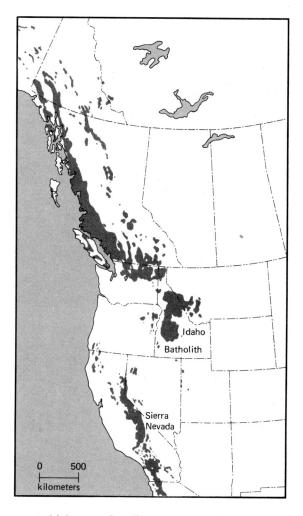

present California coast, a great thickness of sediment accumulated as deep sea fans, fed by submarine canyons at the edge of the forearc basin. These sedimentary rocks are now known as the Great Valley Sequence in central and northern California and their equivalents are found offshore in the southern California continental borderland and in Baja California. Uplift due to the buckling and overthrusting of slices of oceanic crust and sediment broken or scraped off the descending plate created a high area between the forearc basin and the trench, similar to the trench slope break of the modern Sunda Arc in Indonesia (Fig. 5-13). As deformation continued, basins formed between overthrust crustal slices on the trench side of this high area and were filled with normal marine sediments as well as slump debris. This complex mixture of deformed sedimentary and oceanic crustal rocks is known as the *Franciscan Formation* in northern and central California. Similar rocks of Mesozoic age

FIG. 5-10 Yosemite Valley in the Sierra Nevada was carved from Mesozoic granitic batholiths by Pleistocene glaciers. (W. C. Bradley)

FIG. 5-11 The Cretaceous Mesocordilleran Highland was an important source area for sediments deposited in the oceans adjacent to it.

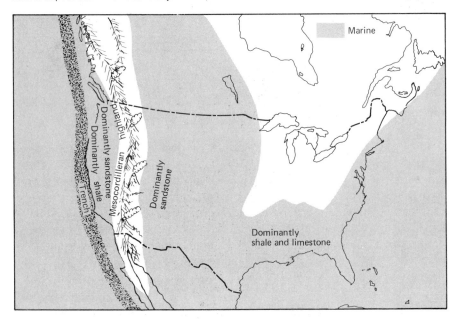

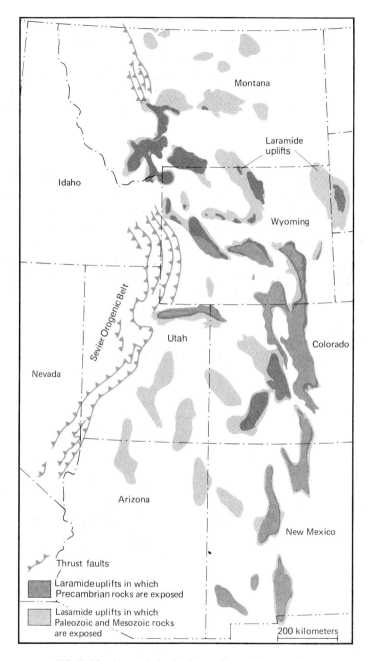

FIG. 5-12 Areas of the Sevier and Laramide orogenies.

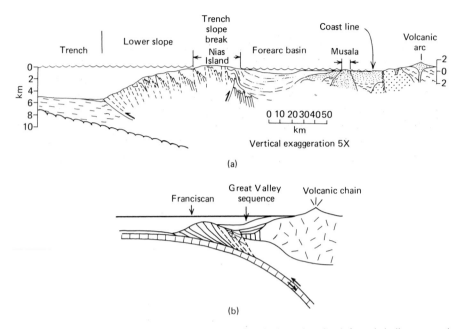

FIG. 5-13 (a) Cross section across the modern Sunda Arc, showing inferred shallow crustal structure. (b) Reconstruction of Late Cretaceous arc-trench system in California. (Karig and others, 1979; Sealy, 1979)

are found along the western margin of much of North America. They record several episodes of subduction and also of accretion of volcanic island arcs and small sialic plates or *microcontinents* onto the western edge of the continent. These island arcs and microcontinents, which were formed somewhere else and later plastered onto a continent, are called *exotic terranes*. Westernmost North America appears to be a mosaic of accreted terranes of both Paleozoic and Mesozoic age (Fig. 5-14), but most of the best-documented ones were accreted during the Mesozoic. Paleomagnetic data indicate that the rocks in the accreted terranes came from far to the south, many from the southern hemisphere!

The western margin of South America had a very similar history to that of North America. The first period of major deformation and volcanism occurred during the Late Triassic, possibly as a byproduct of the initial stages of separation of North America from Gondwanaland. A second deformation in the Late Jurassic may have been related to an increase in subduction activity as South America began to separate from Africa. The most intense episode of deformation began late in the Early Cretaceous and culminated in the Late Cretaceous and early Cenozoic. During this episode huge batholiths were emplaced that today make up much of the Andean mountain chain.

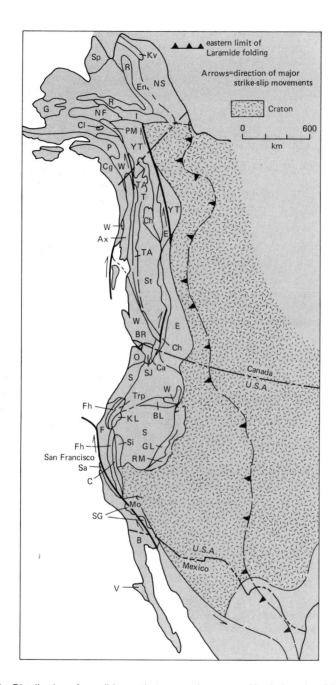

FIG. 5-14 Distribution of possible exotic terranes in western North America. Most (but not all) appear to have accreted during the Mesozoic. The letter designations refer to the names of individual terranes. (Beck, Cox, and Jones, 1980)

133

MESOZOIC ENVIRONMENTS

Red Beds

Red beds occur throughout the geologic record from the middle part of the Precambrian to the Quaternary, but in strata of latest Paleozoic and early Mesozoic age red sandstones, shales, and conglomerates of terrestrial and marginal marine origin are particularly widespread throughout the world. In western North America large quantities of Triassic sediment were transported westward across vast alluvial plains to the Cordilleran Sea. Strata, now red, accumulated on this plain and in bordering tidal flats and lagoons and they lend much of the spectacular color to the canyon country of the Colorado Plateau (See Fig. 1-2).

The red coloring in red beds is caused by ferric iron oxide, usually the mineral hematite, and surprisingly little iron can color a large volume of rock. (Some bright red rocks contain as little as 0.1 percent hematite.) Once formed, ferric iron oxide is very stable as long as the chemical environment is oxidizing. In areas where the groundwaters are alkaline, as they generally are in arid regions, hematite is stable even in somewhat reducing environments.

In the past most geologists thought that red beds formed directly from the red detritus eroded from areas of lateritic soils. Laterites are striking red soils, which, like red beds, are colored by hematite pigment. Because laterites are found only in tropical regions of high rainfall, red beds were widely assumed to be indicators of similar climates in the geologic past. Many red bed sequences, however, also contain beds of gypsum or other evaporites, which form in arid rather than humid climates. Moreover, stream sediment carried from lateritic areas in Central and South America is seldom red. Usually it is yellow or brown. This fact suggests that the red iron oxides in the lateritic soils are either reduced in the fluvial environments or masked by larger volumes of nonred sediments. Microscopic studies of thin-sections of ancient sandstones show that the red ferric iron oxide is largely diagenetic in origin; that is, it was not present in the sediment when it was deposited, but instead it formed later from the chemical alteration of dark, iron-bearing mineral grains. The red pigment in these sandstones is not uniformly distributed throughout but is concentrated around partially etched and oxidized iron-bearing grains.

Red beds are not indicators of any one particular climate. In order for red beds to form, unstable iron-bearing grains must be present and the chemical environment within the deposit must promote the alteration of these grains and the oxidation of iron. Interpretations of paleoclimates must rely on other types of evidence, such as associated sediments and fossils. The red beds of the late Paleozoic and early Mesozoic coastal plains of North and South America are associated with evaporites and probably indicate arid climates. Other ancient red beds associated with fossil tropical floras or coal beds probably formed in humid regions. Thus the formation of red beds does not require a special climate in the source area but rather a certain chemical environment within the depositional basin.

Sand Dune Deposits

Early in the Jurassic Period in the western United States the Navajo Sandstone, a widespread blanket of well-sorted, highly cross-stratified quartz sand more than 300 meters thick, was deposited over a large region of the Colorado Plateau. The near absence of minerals other than quartz and the roundness of these grains suggest that the sand was recycled; that is, it was derived from still older sedimentary rocks. The Navajo Sandstone consists of spectacular, steeply dipping cross beds that occur in sets that are typically up to 15 meters in thickness. The cross-bed directions indicate transportation from north to south (Fig. 5-15) and, indeed, far to the north in Montana and southern Canada there is a widespread unconformity overlain by Middle Jurassic strata, suggesting that upper Paleozoic and Triassic strata were being eroded

FIG. 5-15 The Navajo Sandstone probably formed from marginal coastal dunes inland from the Early Jurassic sea.

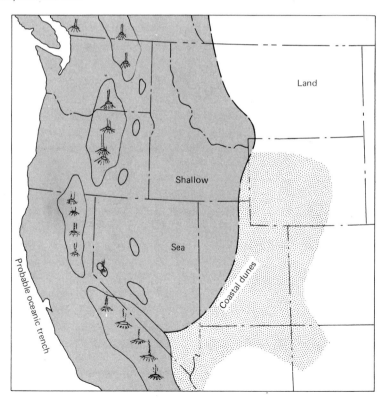

from this region during the deposition of the Navajo. They may well have served as the source area.

The Navajo is only one of many sandstone deposits that have huge sets of tabular cross beds and that are clearly of eolian (wind-blown) origin. In the Colorado Plateau region similar sandstones occur in both the underlying Permian and Triassic and in overlying Jurassic strata. Elsewhere in the world eolian sandstones are known from virtually every part of the geologic column.

The large-scale, steep (up to about 35°) cross bedding that is a famous trademark of the Navajo Sandstone is perhaps the best evidence for its eolian origin, but numerous other kinds of evidence exist as well (Fig. 5-16). The Navajo contains contorted beds that were produced by the slumping and crumpling of unconsolidated sand beneath the enormous load of what must have been huge, rapidly deposited dunes. Although common in dune sands, contorted beds are not restricted to them but occur in some sands deposited by subaqueous processes as well. Other features that are even more indicative of an eolian origin for sands like the Navajo include distinctive ripple marks that are straight crested and have a very low amplitude relative to their wavelength. Unlike ripples formed by water waves, eolian ripples are commonly aligned down the slope of ancient cross beds rather than horizontally across the slope. In addition, sandstones of eolian origin generally contain raindrop impressions and footprints on bedding surfaces, plus distinctive layers of coarse, granule-sized grains in layers that are interpreted as lag deposits. That is, the granules were concentrated by being left behind when finer sand grains were

FIG. 5-16 Eolian crossbedding in the Navajo Sandstone, southern Utah.

blown away. Some eolian sandstones even contain occasional ventifacts — pebbles that are distinctively faceted, similar to those produced today by wind sandblast on rocky desert floors. The Navajo must have been deposited as a huge region of large dunes. Yet instead of a vast Saharalike desert, the Navajo appears to represent a field of coastal dunes that were transported inland from beaches and bars along the Early Jurassic shoreline that lay to the west (Fig. 5-15).

Deltas

In the western interior of the United States, where Cretaceous sedimentary rocks are widespread, the greatest thicknesses of strata accumulated close to the fluctuating western shoreline of the sea. Many of these deposits consist of intertonguing marine and marginal marine sedimentary rocks that include thick, extensive coal beds. The coals are the product of peat deposition in vast coastal marshes, some behind extensive barrier islands and others on large aggrading delta plains.

Most sediments deposited in the sea and along the shoreline of the sea are transported there by rivers. At the point where a river flows into the sea, its velocity is checked; its transporting capacity ceases, and most of its sedimentary load is deposited on the spot. These sediments are then available for transportation out to sea and laterally along the shoreline by ocean waves and currents. Where a river's deposits build into the sea faster than the waves and currents can remove them, the shoreline locally advances seaward and the bulge of sediment thus formed is called a *delta*. Deltas grow at the mouth of a river only if the rate of influx of sediment exceeds the rate of removal by marine processes.

As a delta grows seaward, the river must flow progressively farther across the delta plain to reach the sea. Sediments deposited on the delta plain, along the stream courses and in adjacent marshes and bays, are called *topset beds* (Fig. 5-17). Most of the sediment is delivered to the delta margin. Here much of the sand fraction is distributed laterally along the shoreline by marine currents and deposited as beaches and bars. Most of the silt- and clay-sized fraction continues seaward beyond the river mouth, suspended in the river water plume, which, being lighter than the clear marine water beyond, flows outward as a surficial layer. The fine materials soon settle in the deeper water not far offshore where they become part of the prodelta silty clay. The sediments that collect on the sloping subaqueous portion of the delta are called *foreset beds*. As the delta grows seaward, the delta front sands and other kinds of topset beds are built on top of the relatively homogeneous marine foreset beds, which, in turn, overlie the bottomset beds of the basin floor. The resulting stratigraphic sequence is like that shown in Fig. 5-17. All deltas are crossed by *distributaries*, branches of the river that radiate from the main channel like the spokes of a wheel and discharge at several localities on the delta margin

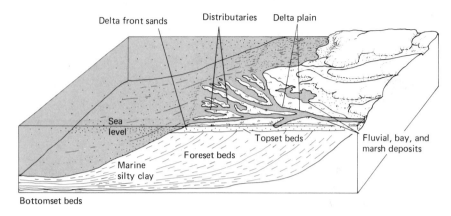

FIG. 5-17 Cross section through a growing delta. Vertical exaggeration is 50 times.

(Fig. 5-17). Marshes abound between the distributaries. With subsidence, thick beds of peat may accumulate from them. Coal beds are common in deltaic deposits preserved in the geologic record.

As the delta extends seaward, the upstream portion of the delta plain aggrades and progressively shorter, steeper courses to the sea become available to the river. Finally, the river abandons its channel and takes a new course to the sea and a new delta lobe is built adjacent to the old one, which sinks entirely below sea level as the area continues to subside. Eventually when the river again shifts its course, a still newer delta lobe may again build outward and over the sunken portion of the original delta lobe. Individual lobes may be built quickly in large deltas. Seven such delta lobes form the modern Mississippi Delta and the entire complex has been built in the last 5000 years (see Fig. 5-18). Ancient delta systems in the stratigraphic record thus consist of intertonguing marine and nonmarine sedimentary rocks. In many of these the well-sorted sandstones deposited by waves and currents along the front of a delta make excellent petroleum reservoirs.

Barrier Islands and Tidal Marshes

The Mesocordilleran Highland furnished a large supply of sediment to the western interior seaway throughout most of the Cretaceous Period. Some of the sand that was brought to the sea by streams was transported laterally along shore by waves and currents and finally deposited in elongate barrier islands similar to those that parallel the Texas Gulf Coast shoreline today (Fig. 5-19). Most of the mud that was brought to the sea was deposited in the deeper waters that lay seaward, to the east of these barrier islands. To the west between the barrier islands and the coastal plain lay vast areas of marshes, tidal

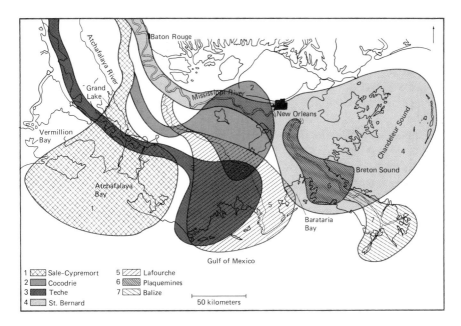

FIG. 5-18 Overlapping subdelta lobes of the modern Mississippi delta complex during the last 5000 years. The site of major deltaic sedimentation has shifted numerous times. (Kolb and van Lopik, 1966)

flats, and lagoons, which were protected from the wave energy of the open sea by the barrier islands themselves. Extensive peat deposits, which were later to become coals, formed in the quiet marshes behind the barrier islands. Because the entire region was continually subsiding, the deposits of peat commonly attained thicknesses of many meters. The barrier islands that flanked portions of

FIG. 5-19 Barrier island–tidal marsh depositional model for Cretaceous coal deposits in the western interior of the United States. Vertical exaggeration 200 times.

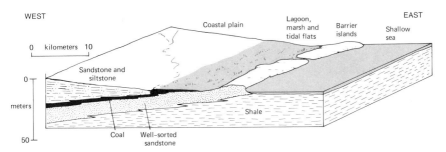

the interior Cretaceous shoreline were built, buried, and rebuilt many times throughout the Cretaceous Period. Individual barrier islands tended not to remain stationary but to prograde seaward; that is, they built seaward over former offshore muds in areas where the rate of deposition exceeded the rate of subsidence (Fig. 5-19). The protected marshes and associated environments behind the barriers similarly shifted progressively seaward and gradually covered former barrier island deposits. In this way, the complex of shallow-water, nearshore sediments, including barrier island, marsh, and lagoonal deposits, became widespread. A marsh that is no more than 10 kilometers wide at any one time may, by the mechanism of seaward progradation, give rise to a coal bed that is several tens of kilometers wide (Fig. 5-20). Following each eastward progradation, the shoreline again moved westward and buried the complex of barrier, marsh, and lagoonal sediments with marine transgressive deposits. Repeated regressions and transgressions of the sea produced thick sequences of alternating marine and nearshore sediments with their associated coal beds. The Cretaceous coal beds produced during these shoreline fluctuations in back-barrier environments, and on nearby delta plains, constitute most of the United States' considerable coal reserves. These coals occur in a broad belt from the Four Corners region northward to Montana (Fig. 5-21).

FIG. 5-20 Cretaceous and early Cenozoic coal fields of the western United States. (U. S. Bureau of Mines)

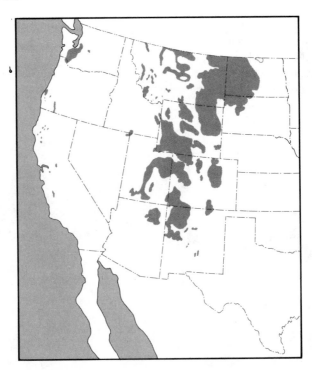

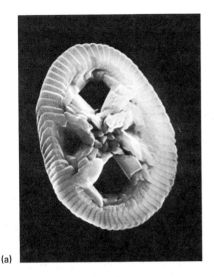
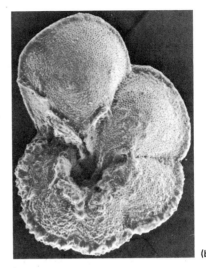

(a)   (b)

FIG. 5-21 Mesozoic calcareous plankton. (a) A Cretaceous coccolithophor, × 10,000 (Courtesy of Hans Thiersten); (b) A Cretaceous planktonic foraminifer, × 80 (Courtesy of R. Mark Leckie).

Calcareous Ooze

At the same time that vast shallow seas occupied large portions of the Mesozoic continents and produced extensive marginal marine and shallow marine deposits, sediments were also accumulating in the deep ocean basins. Jurassic and Cretaceous sedimentary rocks in the deep sea provide the first clear views of undisturbed open oceanic environments. Nowhere has any portion of the deep ocean basins been found that is older than the Jurassic. The likelihood of finding any in the future seems small. Most ocean floor older than Jurassic has been subducted and thus returned to the mantle, although a small portion has been accreted to the continents as ophiolite rocks in mobile belts.

During the Jurassic Period, shortly after the modern ocean basins had begun to form, major evolutionary events affected the open ocean's planktonic plants and animals, which, in turn, profoundly altered the worldwide pattern of carbonate sedimentation. Prior to this time there were no important carbonate-secreting planktonic organisms and almost all the calcium carbonate removed from the oceans had been precipitated by benthic organisms on continental shelves and in epicontinental seas. In these shallow settings the resulting limestones were recycled readily by frequent uplift and erosion. Then in the Jurassic the evolution of tiny calcareous planktonic marine algae called *coccolithophorids* [Fig. 5-21(a)] and calcareous *planktonic foraminifera* [Fig. 5-21(b)] caused large quantities of tiny calcium carbonate skeletons to be produced in the upper photic portions of the deep oceans. The rain of these tiny particles onto the sea bottom produced the first deep-sea calcareous oozes. Although subject to dissolution, particularly at great depths, calcareous ooze is not readily eroded by mechanical means and is immune to small changes in sea level and minor uplifts that serve to erode rocks deposited at the continental margins. With burial, calcareous ooze becomes lithified to a distinctive soft, limestone called *chalk*. Chalk has a unique texture because it is a weakly cemented and highly porous agglomeration, predominantly of coccoliths and foraminifera. Planktonic foraminifera are more soluble than coccoliths and are preferentially dissolved during the processes of compaction and cementation. As a result, the proportion of coccoliths tends to be much higher in chalks than in unindurated oozes.

Deep-sea drilling has revealed extensive chalk deposits of Cretaceous and Cenozoic ages in the deep ocean basins and clearly calcareous ooze has been the most widespread kind of sediment deposited in the deep sea since the Jurassic. Today calcareous ooze covers nearly 50 percent of the deep-sea floor. Although coccoliths and planktonic foraminifera are also deposited extensively on continental shelves, here they do not form oozes because they are masked by terrigenous material and the remains of benthic organisms. During the Cretaceous, however, chalks now exposed on the continents were produced in water only a few hundred meters deep at most, judging by the benthic organisms they contain. A well-known example is the chalk that forms the famous white cliffs on the southern coast of England. These chalks typically contain numerous phosphatic nodules, which are evidence of very low rates of sedimentation. Thus they represent oozes that accumulated slowly and with almost no dilution by terrigenous sediment. The Cretaceous epicontinental seas must have been vast and the areas of chalk deposition far removed from terrigenous source areas at times of maximum transgression.

Sea Level Changes

What caused the numerous transgressions that created the widespread epicontinental seas of the Jurassic and the Cretaceous? Many transgressive-regressive cycles of epicontinental seas are, in fact, comparatively small in scale and reflect changes in sea level of only a few meters. These small fluctuations in sea level may be only local in an actively subsiding depositional basin; that is, a portion of a continent may temporarily rise or sink as a result of a change in the rate of subsidence or the rate of sedimentation. These are only changes in *relative* sea level. The handful of truly major transgressions and regressions, however, affected not only entire depositional basins, such as the western interior seaway in North America, but they also have simultaneous counterparts on other continents. Thus they almost certainly resulted from large worldwide changes in sea level. A real change in sea level that affects all oceans simultaneously is called a *eustatic* change. On a still larger scale, the Jurassic and Cretaceous periods can be viewed as a single huge cycle of eustatic rise and fall of sea level during which all continents were inundated by the ocean (Fig. 5-22).

Eustatic sea level changes, which are worldwide by definition, may be a

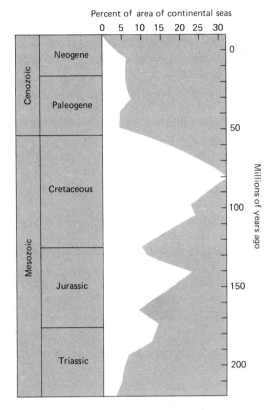

FIG. 5-22 Proportion of North America covered by epicontinental seas in the Mesozoic and Cenozoic. (A.G. Johnson, 1971)

result of (1) changing the quantity of water in the oceans or (2) changing the capacity of the ocean basins to hold the water they have. During the Pleistocene ice ages eustatic sea level changes occurred because of the first reason. When ice accumulated periodically on the continents, sea level dropped, by 200 meters or so due to withdrawal of water. During interglacial times sea level rose again. Shorelines on some coasts moved landward hundreds of kilometers. This mechanism, however, does not appear to be adequate to explain the several hundred meters of eustatic sea level change inferred for the Cretaceous Period. Moreover, there is no direct evidence for continental glaciation anywhere in the world during the Cretaceous.

Instead the large eustatic sea level fluctuations during the Cretaceous probably resulted from changes in the actual capacity of the ocean basins — changes that may have been partly caused by changes in the rate of sea floor spreading. After being generated at an oceanic ridge, new oceanic crust subsides at a uniform rate while simultaneously being transported away from the ridge as part of diverging crustal plates. The subsidence of the new oceanic crust is a function of time and not of distance from the ridge. The ridge flanks are comparatively steep where the spreading rate is slow and gentle where the spreading rate is fast. Thus, if spreading is slow, the resulting narrow ridge leaves a great deal of room for water in the ocean basins. When spreading rates increase worldwide, however, water is displaced by the broadening ridge, sea level rises, and the oceans spill onto the continents in marine transgression. Thus a possible mechanism for the large-scale, period-long worldwide Jurassic and Cretaceous transgressive-regressive marine cycles is worldwide increases, followed by decreases, in sea floor spreading rates. Yet whether the Jurassic and Cretaceous Periods as a whole were in fact times of accelerated sea floor spreading is not known. Even if they were, it seems unlikely that changes in spreading rates could have been rapid enough to account for the numerous short-term eustatic fluctuations that occurred. The explanation for the Jurassic and Cretaceous eustatic sea level changes may lie in some other causes that are not yet understood.

THE MESOZOIC EXPANSION OF LIFE

Modernization of the Living World

The severe reductions of the living world at the end of the Permian were not followed immediately by the appearance of new and varied faunas and floras, but as the Mesozoic Era progressed, new evolutionary radiations ultimately led to a far greater diversity of organisms than existed at any time during the Paleozoic Era. In the ocean many new invertebrate groups arose during the early part of the Mesozoic to replace those eliminated in the late Paleozoic extinctions. The earliest fossiliferous Triassic strata contain a few conservative species of ammonites, although they are locally represented by large numbers of individuals. As the Triassic Period progressed, the ammonites diversified and filled a wide range of environments. Some 400 genera belonging to ten families evolved. For unknown reasons, however, most became extinct before the close of the period. Only a single family of ammonites survived the Triassic to give rise to a vast and complex group of 1200 genera that dominated the Jurassic and Cretaceous seas. So rapid was the evolution and extinction of genera and species throughout the Mesozoic that ammonites serve as the best guide fossil known for worldwide correlation.

Other mollusks show a different evolutionary pattern. Clams and snails assumed a modern appearance by the gradual addition of new families, many of which are still living today. Clams evolved the capacity to burrow in soft sediments and these groups diversified quickly. Other invertebrate phyla that gradually attained a modern appearance during the Mesozoic were the corals, crustaceans, and echinoderms, which continue as the dominant invertebrates of oceans today.

The tabulate and rugosan corals, which flourished during the Paleozoic, died out at the end of the Permian. Triassic corals belong to the Scleractinia, the same corals we have today, but they were not common in carbonate build-ups of the Middle Triassic. Instead large sponges and encrusting tubular organisms continued to be important, as in the Permian. Finally, in the Late Triassic, reefs took on a modern aspect as large colonial scleractinian corals became the dominant reef builders. Except for certain Cretaceous buildups of some rather bizarre bivalves, known as rudists, reefs have been basically similar in structure and fauna since the Late Triassic.

Figure 5-23 shows an example of the internal configuration of a Jurassic reef. The vertical sequence of corals inhabiting the Jurassic reef as it built upward into very shallow water is similar to the depth distribution of growth forms on modern reefs. The corals in the basal, early stage of the Jurassic reef were platy, sheetlike forms similar to those found in the deeper parts of modern reefs, where little sunlight penetrates. Reef-building scleractinian corals contain symbiotic algae that require light for photosynthesis. The platy form of deep reef corals enhances the surface area that is exposed to light. More erect, branching forms superceded the flat shapes and head corals superceded the branching forms in the developing Jurassic reef—just as they do in progressively shallower areas of modern reefs. Both Jurassic and modern reefs are capped by encrusting corals and red algae.

FIG. 5-23 Idealized Upper Jurassic reef. Upward changes in coral growth forms from sheety to dendroid to massive encrusting, and from sponges to corals to red algae as the reef developed, are similar to changes from deep to shallow parts of modern reefs. (Wilson, 1975)

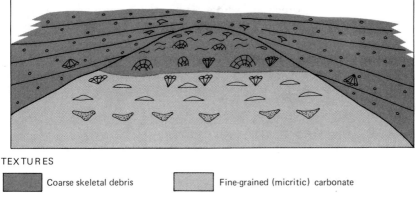

TEXTURES

☐ Coarse skeletal debris   ☐ Fine-grained (micritic) carbonate

DOMINANT FOSSILS

~~ Encrusting corals & red algae   ⊕ Head-shaped corals

▽ Fasciculate (tube) corals   ⌒ Sheet & platy corals

⌁ Brachiopods & ancestral rudists   ▽ Sponges   ∘∘∘ Rounded skeletal debris

Bryozoans, brachiopods, and crinoids were much less abundant in Mesozoic reefs than in the Paleozoic and mollusks, especially bivalves, were much more important. The rudist bivalves evolved in the Late Jurassic and were important in some Cretaceous carbonate buildups. These bivalves developed one very large valve and one small one that served much like a lid (Fig. 5-24). Although rudists were not colonial and so did not build a true framework, they grew close together in such intertwined masses that their bulk made substantial buildups (Fig. 5-25). The mass of rudists acted as a baffle—enhancing the accumulation of fine-grained sediments—and provided a habitat for other reef-dwelling organisms. Stromatoporoids, were also important in Mesozoic reefs, both as frame builders and encrusters, but along with the rudists, they became extinct at the end of the Cretaceous. Since that time reefs have been dominated by large frame-building scleractinian corals and a great variety of red and green algae, which bind and encrust the reef and also contribute large quantities of both coarse and fine sediment to the bulk of the reef.

Paralleling the modernization of marine benthic life were changes in the plankton. Besides the appearance of the coccolithophorids and planktonic foraminifera, major changes also took place among the silica-secreting plankton: diatoms appeared and many new kinds of radiolarians evolved. By the end of the Mesozoic life in the oceans was, in most respects, closely similar to life as it exists today.

Perhaps the most dramatic biotic changes of Mesozoic time, at least from our point of view as land dwellers, took place on the continents where new groups of land plants and animals developed. The fossil record shows that the land plants of Paleozoic time were mostly seedless, fernlike forms as well

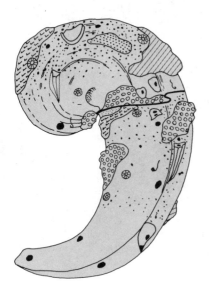

FIG. 5-24 A rudist bivalve showing the unequal valves. This individual is heavily encrusted with a variety of other organisms. (Kauffman, 1973)

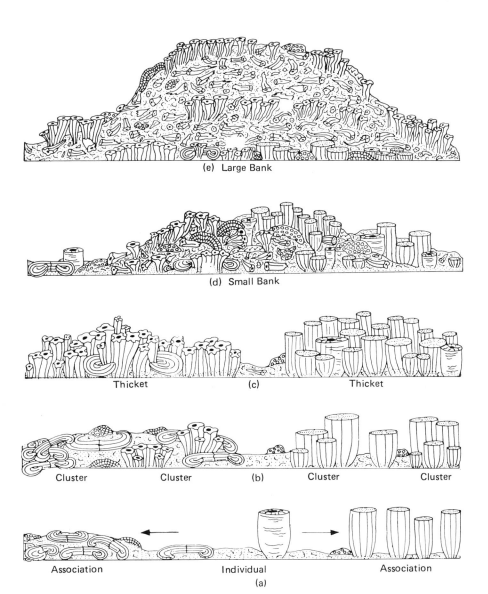

FIG. 5-25 Growth stages in the formation of a rudist carbonate buildup. (a) Associations of individual rudists colonize a shallow part of the sea floor. Shells accumulate and develop small lenticular frameworks called (b) clusters. These, in turn, merge into larger aggregations called (c) thickets, which coalesce into (d) small, lens-shaped banks. These grow upward into (e) large banks that contain a maximum diversity of rudists. (Kauffman, 1973)

147

as primitive seed-bearing ancestors of modern conifers. These groups continued as the principal land plants until Early Cretaceous time when the more advanced flowering plants rapidly expanded to dominance over much of the land surface.

At the end of the Paleozoic the mammallike reptiles were the dominant ones on earth. There were more than ten times as many mammallike reptile genera as other kinds of reptiles. This situation was dramatically reversed in the Triassic. The thecodonts, the ancestors of dinosaurs, were represented by about five times as many genera as the mammallike reptiles. The Mesozoic was truly the Age of Reptiles. Flying reptiles evolved and became huge. A variety of marine reptiles, such as ichthyosaurs and plesiosaurs, flourished in the sea. The first representatives of our modern reptiles — turtles, crocodiles, snakes, and lizards — appeared and dinosaurs ruled the land. Mammals evolved in the Triassic, but they remained small and were never abundant during the Mesozoic.

Warmblooded Dinosaurs?

Today mammals are the dominant terrestrial vertebrates and have been since the demise of dinosaurs during the major extinction event at the end of the Cretaceous. Why have mammals become the dominant group since dinosaurs died out? What is it about mammals that has apparently made them competitively superior to the remaining reptiles?

The most frequent answer is that they are warmblooded. They maintain a relatively constant internal body temperature and thus can remain active over a much larger temperature range than exothermic (or coldblooded) organisms. There is another advantage to endothermy: much greater endurance. The metabolic processes that maintain a constant internal body temperature also sustain a high level of activity over long periods of time. Moreover, the four-chambered heart found in birds and mammals provides a more efficient delivery system for oxygenated blood. Both reptiles and mammals are capable of running at high speeds over short distances, but a mammal can keep running for a much longer time. If mammals can inhabit a much wider range of temperatures and have greater endurance than reptiles, why is it that they did not become the dominant terrestrial vertebrates soon after evolving in the Triassic? Why did they diversify and become abundant only after dinosaurs died out?

Such questions have led a number of paleontologists to conclude that dinosaurs were also warmblooded. Is there any evidence to support this conclusion? The answer seems to be "perhaps." Internal maintenance of a constant body temperature requires about ten times more food. Consequently, mammalian communities have low predator–prey ratios: a few predators require a large number of prey. Estimates of the abundance of predator and prey species suggest that most late Paleozoic reptile communities contained a relatively large number of predators (*high* predator to prey ratio) compared to

Tertiary mammalian communities. Permian communities, however, dominated by mammallike reptiles had *low* ratios (fewer predators), similar to mammalian communities. And where do dinosaurs fit in this picture? Predator–prey ratios for dinosaur communities are similar to those of mammals. This fact suggests that dinosaur energy requirements were high, as they are for endothermic mammals.

Evidence from the internal bone structure of dinosaurs also suggests that these reptiles may have been endothermic. The bones of mammals have a spongy interior that is related to the development of blood vessels within the bone. The vessels facilitate the rapid exchange of calcium and (especially) phosphorus between blood and bone. Phosphorus bonds are the major carriers of energy in biological systems and the rapid exchange of phosphorus probably aids in sustaining high levels of activity. The bones of most modern reptiles are solid in the center, but dinosaurs had spongy bone. Reconstructions of dinosaur skeletons indicate that dinosaurs had relatively long-legged upright stances, with the legs directly under the body rather than out to the sides as in most other reptiles. Such findings also suggest that dinosaurs were relatively agile and capable of a high level of activity.

The skeleton of the earliest known bird, *Archeopteryx* (Fig. 5-26), is similar to those of certain small dinosaurs. In fact, if the *Archeopteryx* remains did not include obvious feathers, the skeleton would have been classified as a dinosaur. Because birds are endothermic, it seems possible that their immediate dinosaurian ancestors may also have been warmblooded. The skeletal features of *Archeopteryx* indicate that it was a ground-dwelling predator and did not fly. Small animals have a relatively large surface area-to-volume ratio,

FIG. 5-26 *Archeopteryx*, the first bird. (Courtesy of the American Museum of Natural History)

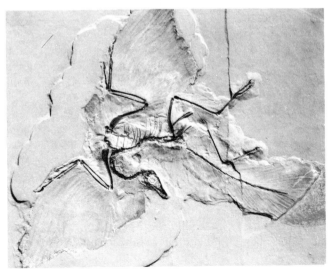

which enhances heat loss, and feathers probably evolved as an adaptation for the insulation of small, endothermic dinosaurs.

The large size of many dinosaurs makes direct comparison of their food requirements, activity levels, metabolic rates, and heat retention with those of modern reptiles and mammals difficult. We do not know whether changes in these variables with body size can be extrapolated correctly to such vastly larger bodies. The dinosaurs' large size may have produced a relatively constant body temperature without internal mechanisms for temperature regulation because a large body has a relatively small surface-to-volume ratio, which reduces heat loss. Maintenance of a relatively stable body temperature would have permitted an active life without internal heat regulation. The very size of dinosaurs suggests that a four-chambered heart may have been necessary to meet their blood circulation requirements, but ectothermic crocodiles also have a four-chambered heart today. Other evidence for endothermy in dinosaurs has been challenged as well and the question may never be settled unequivocally. It does seem likely that many dinosaurs were active animals with a relatively constant body temperature. Some, perhaps most, were probably truly endothermic.

The Cretaceous Extinction Event

The extinctions at the end of the Mesozoic, like those at the end of the Paleozoic, drastically affected marine as well as terrestrial life. In the sea extinction was more widespread among pelagic animals and plants than among bottom-dwelling organisms. The ammonites, which were among the most common Mesozoic marine invertebrates, were wiped out completely and only a sprinkling of the formerly diverse planktonic foraminifera and coccolithophorids survived into the Cenozoic. On land three of the six major reptile groups became extinct (Fig. 5-27).

Some of the same theories used to explain the major Permian extinction have also been invoked for the Cretaceous. The widespread Cretaceous chalk deposits with their abundant phosphate nodules, for instance, are cited as evidence for low-lying continents contributing little terrigenous sediment to the oceans, leading to nutrient depletion, massive extinction in the plankton, and low levels of atmospheric oxygen. The evidence for Cretaceous nutrient depletion and low oxygen is not too compelling. The widespread Late Cretaceous mountain building along the western margin of the Americas, in the North American western interior, in southern Europe, southeast Asia, Antarctica, and Japan does not fit well with the view of low, quiet continents furnishing little nutrient material to the sea. Although the vast majority of Cretaceous coccolith species did not survive into the Cenozoic, there is no evidence that their dramatic decrease in diversity was accompanied by a corresponding decrease in total numbers of individuals and hence by a decrease in the oxygen they produced.

Several theories involving increased radiation from space have been pro-

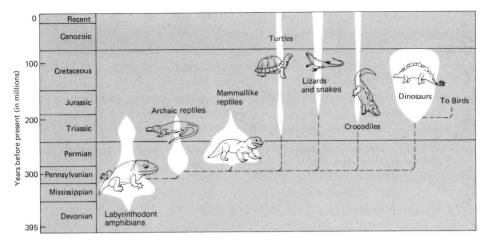

FIG. 5-27 Evolutionary history of the reptiles. Dashed lines show the most probable evolutionary relations of the groups. The width of the vertical areas indicates the approximate abundance of each group.

posed. Usually they involve either the postulated occurrence of a supernova in a nearby part of our galaxy or a reduction in protection from radiation as the intensity of the earth's magnetic field declines drastically during a polarity reversal. The principal difficulty with the increased radiation theories is that marine organisms should have been much less affected than terrestrial organisms because even a thin layer of water is a very effective radiation shield. Although it is true that dinosaurs died out, other terrestrial life was not drastically affected and the most striking extinctions took place in the sea.

Low-salinity surface waters in the world ocean have been suggested as a reason for the Cretaceous extinction as well as for those of the Permian, although the proposed origin of the low-salinity water is quite different. The theory proposes that during the Cretaceous the Arctic Ocean became cut off from the rest of the world ocean and was a brackish, possibly even fresh, body of water. Rifting of Greenland from Norway opened a connection between the North Atlantic and Arctic oceans, thus allowing the freshwater to spill out at the surface as higher-salinity Atlantic water entered along the bottom. The resulting low-salinity layer, which is estimated to have been several tens of meters thick, killed off the plankton and prevented mixing of oxygen into the deeper layers, thereby causing the ammonites to become extinct as well. This cool Arctic Ocean water is also supposed to have been responsible for lower temperatures, less precipitation, and a very short-lived (10 years) but severe drought that caused the extinction of the dinosaurs. There is no evidence for a fresh or brackish Arctic Ocean in the Late Cretaceous and the proposed combination of low oxygen and low salinity surely would have had a devastating effect on Cretaceous shelf benthos. Yet benthic organisms were relatively unaffected by the Cretaceous extinction event. The late Maastrichtian timing of the rifting of Greenland from Norway is also open to question.

The discovery of unusually high concentrations of iridium in sediments at the Cretaceous–Tertiary boundary in such widely separated places as Denmark, Italy, the North Pacific Ocean, and New Zealand has led to speculation that the impact of an asteroid may have been responsible for the Cretaceous extinctions. Relative abundances of iridium and other rare metals in the boundary sediments are similar to those found in stony meteorites and dissimilar to those in common earth rocks. According to this theory, the impact of the asteroid caused vast quantities of dust to be thrown up into the stratosphere, where it was carried around the entire earth, causing darkness for several years. Without sunlight, the phytoplankton was drastically reduced, taking the higher levels of the planktonic and nektonic food chain with it. Presumably benthic organisms fared better because they could feed on detritus. Although the evidence for an extraterrestrial source for the iridium is strong, there seems to be considerable question whether a period of several years' darkness could explain the observed patterns of extinction, especially among land plants.

There is some evidence from oxygen isotope data that temperatures declined near the end of the Mesozoic. This factor may have contributed to the demise of dinosaurs, especially if they were endothermic. Endothermic mammals and birds maintain a high internal temperature despite low environmental temperatures by growing a protective layer of insulation (hair or feathers), but reptiles do not have this capability. Low temperatures cannot have been the major factor in the extinction of Cretaceous plankton and the ammonites, for the considerably more profound decreases in global temperature during the Pleistocene did not cause significant extinctions in the marine realm.

six

the cenozoic earth

PLATE TECTONICS: INFLUENCE ON GEOGRAPHY

The Global Picture

By the beginning of the Cenozoic Era, about 65 million years ago (Fig. 6-1), the continents had begun to achieve their modern shapes. The Atlantic Ocean continued to widen at the expense of the Pacific, and the Indian Ocean continued to open as India drifted northward (see Fig. 5-4). Precisely when India first collided with Asia is not yet known, but it may have been as early as middle Eocene. Uplift of the Himalayas at the former Asian continental margin began in the Oligocene and their maximum deformation took place in the Pliocene and Pleistocene when the northern edge of the Indian continental mass was actually pushed beneath the southern margin of the Asian continent. The result was an extraordinarily thick continental crust that deformed into the highest mountain range on earth.

Australia and Antarctica rifted apart in the Late Cretaceous, about 100 million years ago. By Eocene time, Australia had moved northward and Antarctica southward over the South Pole. This set the stage for the deep circumpolar circulation of the Southern Ocean when the Drake Passage later opened between South America and Antarctica. Also in the Eocene, rifting in the North Atlantic shifted from the west side to the east side of Greenland. Subsequently the Norwegian Sea opened, completing the separation of Europe and North America and producing a deep passage between the North Atlantic and the Arctic. This permitted cold, deep water to flow into the North Atlantic. These circulation changes were probably key factors in the development of the Late Tertiary and Quaternary Glacial cycles.

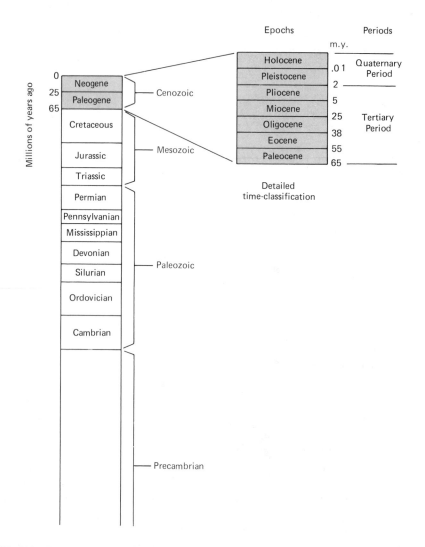

FIG. 6-1 Position of the Cenozoic Era in the geologic time scale. The Cenozoic is divided into periods in two ways. Both are widely used, as are the epoch names.

Early in the Cenozoic the African and European Plates converged and closed the Tethys Sea. In the Oligocene the colliding plates initiated the Alpine Orogeny which built the Alps and Carpathians in southern Europe and the Atlas in north Africa. Small–scale separations of microplates simultaneously produced the Mediterranean Sea behind the collision zone.

The Drying Up of the Mediterranean Sea

The connection between the Mediterranean and the Atlantic began to close in the middle Miocene by the joining of Europe and Africa at Gibraltar. The results were spectacular. In latest Miocene time the Mediterranean, which is 3000 meters deep, dried up (Fig. 6-2). This surprising discovery was made

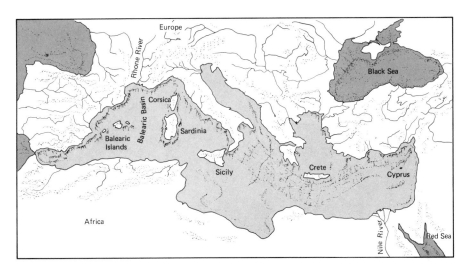

FIG. 6-2 The dried-out Mediterranean Sea is represented by this panoramic drawing of the modern submarine topography of the Mediterranean basin. Approximately 6 million years ago gravels and silts were being deposited around the edge of the basin at the foot of the steep slope and the Balearic Basin was a salt lake where evaporite minerals were being precipitated. (Hsü, 1972)

during drilling in the bottom of the Mediterranean by the Deep Sea Drilling Project. Beneath the Pleistocene and Pliocene sediments on the deep floor of the Mediterranean are widespread beds of anhydrite and halite, which locally reach thicknesses of 1500 meters. Anhydrite and halite are evaporites; that is, they are precipitated from seawater when it is concentrated by evaporation. When they were first discovered, it was thought possible that the Mediterranean had remained deep in the latest Miocene, much as it is today, even as the evaporites formed. The detailed features of these deposits do not appear to be of deep-water origin, however. They contain both nodular anhydrite and stromatolites. Stromatolites are formed by blue green algae, which require sunlight and so must be deposited in shallow waters. Most living stromatolites, in fact, occur at shorelines, in supratidal environments. Although nodular anhydrite may form in a number of ways, it seems to be most common in supratidal flat and playa lake environments and requires relatively warm (35°C) temperatures to form. Such temperatures are not likely at a depth of 3000 meters in the sea but are easily achieved on a playa or tidal flat.

Supporting evidence that the sea was absent from the Mediterranean basin comes from adjacent North Africa and southern Europe, where river gorges are incised far below present sea level (Fig. 6-3). The rivers that cut these gorges in latest Miocene time flowed at a level below the present submerged Mediterranean shelf and even well below the upper part of the continental slope, and this situation could only have occurred if the Mediterranean dried up, thus drastically lowering the base level for the streams flowing into the basin. A final line of evidence for the drying up of the Mediterranean

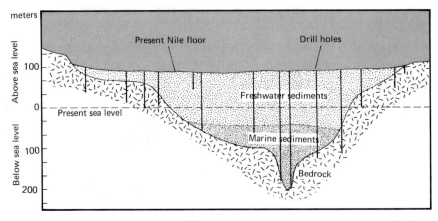

FIG. 6-3 This deep gorge under the upper Nile Valley near Aswan, Egypt, was cut about 200 meters below the present sea level in latest Miocene time when the Mediterranean was dry. When the Mediterranean filled again at the beginning of the Pliocene, the sea invaded this gorge and deposited marine sediments. (Hsü, 1972)

comes from the distribution of the evaporites themselves. On the Mediterranean floor, the evaporitic facies in the Balearic Basin of the western Mediterranean form a bull's-eye pattern with the most soluble minerals in the center. This pattern would be expected from shrinking saline lakes in the present deep depressions on the floor of the Mediterranean and would not be expected in a deep-water evaporite setting.

Actually, it is not too difficult to imagine how the Mediterranean might dry up. At present, only about 10 percent of the water that evaporates from the Mediterranean each year is replenished by rain and by rivers. The other 90 percent flows into the Mediterranean from the Atlantic through the Strait of Gibraltar. If the strait were totally closed now, the evaporation loss would not be replaced and the Mediterranean would dry up completely in about 1000 years. Normal marine sediments interbedded with the evaporites indicate that the Mediterranean basin actually dried up and refilled several times. And, indeed, a single evaporation of all the water in the present Mediterranean would produce less than 10 percent of the 1500 meters of evaporites that were deposited on the bottom of the Mediterranean during the Miocene.

The removal from the ocean of such a large amount of salt may have had a considerable effect on the ocean's salinity and on world climates. Lower salinity facilitates the freezing of seawater and the evaporation of such a large quantity of water in the Mediterranean region must have augmented precipitation elsewhere. Thus the drying up of the Mediterranean may have been an important factor in enhancing the glaciation that took place around the Arctic and on Antarctica during this time.

At the end of the Miocene, about 5.5 million years ago, a rise in sea level, accompanied by tectonic lowering of the Gibraltar sill, allowed the Atlantic Ocean to refill the Mediterranean basin for the last time. Had the water merely trickled in it would have evaporated before it could fill the basin. In order to support even the hardy fauna that appears with the first sediments above the evaporites, the influx would have had to exceed the evaporative loss by a factor of ten, or about 40,000 cubic kilometers of water per year. This is a volume 1000 times greater than that of Niagara Falls, and although the slope over which it flowed was not nearly as steep as the falls, it must have been a spectacular sight to behold!

What of the future Mediterranean Sea? Africa is still pushing toward Europe and the Mediterranean region is tectonically active. Uplift of the 300-meter-deep Gibraltar sill could well cut the Mediterranean off from the Atlantic again at some time in the distant future.

The Joining of the Americas

About 3.5 million years ago, in the late Pliocene, the Isthmus of Panama was uplifted above sea level. This uplift closed the low latitude connection between the Pacific and Atlantic and isolated the marine biotas of the two oceans. The ocean current that had flowed westward from the Atlantic into the Pacific was diverted northward to give additional strength to the Gulf Stream. The warm current pushed far to the north and probably brought increased precipitation to the lands at high latitudes around the North Atlantic, which may have helped to feed the Pleistocene glaciers on land.

The uplift of the Isthmus of Panama provided a land bridge for the interchange of terrestrial mammals between North and South America. The diversity of habitats available on the land bridge must have been greater than at present because the mixed inter-American fauna that occupied both North and South America during the peak of the interchange represents a number of habitats (for example, savannah) that do not exist in the Central American corridor today. Although the number of genera that migrated between the continents was nearly the same in both directions, the impact of the interchange was far more dramatic in the south than in the north. Immigrants from South America occupied less than 10 percent of the North American continent during the height of the interchange and these immigrants had little effect on the native North American fauna. In South America, however, North American mammals ranged over most of the continent; the number of endemic South American genera belonging to the ungulate group alone plummeted from 13 to 3 whereas the number of invading North American genera rose from zero to 14. The impact of the invasion was perpetuated in South America as the new arrivals diversfied into several new lineages while the endemic groups continued to decline.

Cenozoic Tectonics of the United States

During the Mesozoic the Appalachian Mountains on the eastern margin and the Ouachita Mountains on the southern margin of the North American continent were gradually eroded to lowlands. In Cretaceous time the lowlands were actually flooded by transgressing epicontinental seas. Throughout the

Jurassic and Cretaceous sediment derived from the eroding mountains and other sources was deposited on the block-faulted continental margins that had been produced during the Triassic breakup of Pangaea (Fig. 6-4). In the Cenozoic, the thick marginal wedge of sediment continued to grow seaward, finally creating the modern Gulf and Atlantic coastal plains and continental shelves. Broad, gradual upwarping in the late Cenozoic reelevated the Appalachians and rejuvenated the streams flowing across the region. Cretaceous and early Cenozoic strata were eroded from the uplifted area and the streams became en-

FIG. 6-4 The development of the Atlantic continental margin. Stage A shows the Triassic rifting. Stages B, C, and D show subsidence and deposition of the continental shelf sedimentary wedge in the Jurassic, Cretaceous, and Cenozoic. (Schlee and others, 1979)

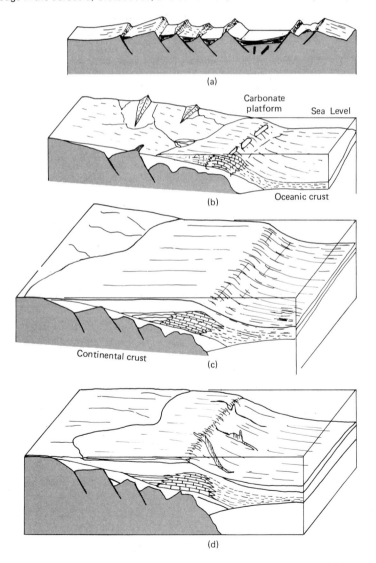

trenched. The Atlantic coastal plain and continental shelf that were gradually built on the east as North America moved westward, away from the spreading mid-Atlantic ridge, characterizes what has been called a "trailing margin".

In contrast, the strong tectonic activity that shaped the western, active margin of the continent during the Mesozoic continued throughout the Cenozoic. In the early part of the Cenozoic the Laramide Orogeny continued to uplift the Rocky Mountain blocks and to depress the intervening basins. Tectonism and, shortly thereafter, volcanic activity extended far inland to encompass much of the western third of the continent (see Fig. 5-11). In the Paleocene fluvial and lacustrine deposits and thick beds of coal accumulated in the Rocky Mountain basins. To the west in Idaho, Nevada, and western Utah, Paleocene deposits are scarce, suggesting that this was a region of broad uplands undergoing erosion. Meanwhile, west of the Sierra Nevada, marine and marginal marine environments prevailed. Relatively small river systems fed deltaic complexes that built westward over a narrow shelf. Submarine canyons supplied sediment to numerous deep-sea fans and to a deep trench that bordered the entire west coast of North America.

In the Eocene and Oligocene broad inland basins received thick sequences of fluvial and lacustrine sediments and many areas of the western United States experienced volcanic activity (Fig. 6-5). One such area was the Cascade Range, which lay inland from its present position. The Cascades have, in fact, rotated about 75° during the last 50 million years to attain their present north-south position. Further inland, large quantities of andesitic volcanics were extruded in the Challis region of Idaho and Montana, the Absarokas of northwestern Wyoming, the Henry, La Sal and Abajo ranges of Utah, and the San Juan Mountains of Colorado (Fig. 6-6).

Subduction of oceanic crust belonging to the Farallon Plate was apparently more rapid than production of new oceanic crust at the spreading center to the west between the Farallon and Pacific plates, for the spreading center reached the trench in the Oligocene (Fig. 6-7). Where the Pacific and North American plates now joined [Fig. 6-7(b)], subduction ceased and relative motion between the plates was taken up by strike-slip movement on the San Andreas transform fault. Subduction continued along the now-separated segments of the trench to the north and south. As more of the Farallon Plate was subducted, the strike-slip boundary between the Pacific and North American plates lengthened [Fig. 6-7(c)]. As the subducting plate boundary changed to a transform fault boundary, volcanism ceased on the continent immediately to the east. Today we can still see this pattern [Fig. 6-7(c)]. The southernmost active volcano of the Cascade Range is Mount Lassen, which lies opposite the end of the San Andreas Fault, where it joins the Mendocino Fracture Zone. The fracture zone forms the boundary between the Pacific Plate and the northern remnant of the Farallon Plate (the present Juan de Fuca Plate). Similarly, there is no volcanism to the south until we reach central Mexico, where a southern remnant of the Farallon Plate (the Cocos Plate) is presently being subducted.

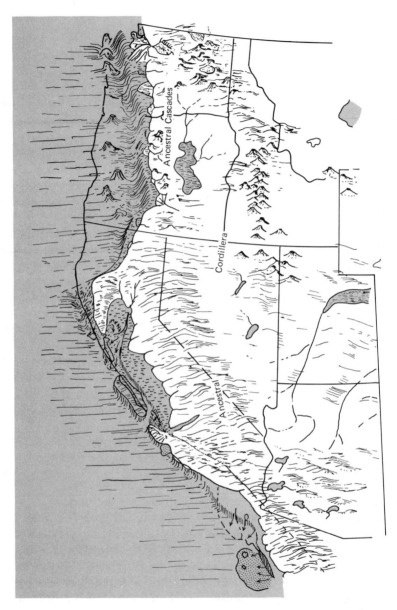

FIG. 6-5 Paleogeographic reconstruction of the western United States during the late Eocene. (Nilsen and McKee, 1979)

    In the Miocene, with the onset of extensional forces, the Great Basin developed its modern "basin and range" character as block faulting created a series of narrow mountain ranges separated by valleys [Fig. 6-8]. With the change from a compressional to a tensional tectonic style, the volcanism changed from dominantly andesitic to basaltic. Several reasons for the onset of regional extension have been proposed. Some geologists have suggested that

FIG. 6-6 The Absaroka Range of northwestern Wyoming is formed entirely of Eocene volcanic rocks.

the previous subducted portion of the Farallon Plate that was no longer being drawn down into the mantle by subduction "floated" upward, doming up the Great Basin area and causing it to be faulted as the crust was stretched over the rising material. Others have pointed out that strike-slip motions, such as that on the San Andreas Fault, generate regional tensional and compressional forces at angles to the shear. Basin and Range faults lie at about the expected angle to the San Andreas Fault and so may have resulted from shearing on the fault as it lengthened with time. Extension behind the rotating Cascade block may also have been in part responsible for the faulting and volcanism in the Great Basin as well as for the extensive Miocene basaltic flows that form the Columbia Plateau (Fig. 6-9). On the Snake River Plain east of the Columbia Plateau volcanic extrusions began in the Miocene in southwestern Idaho and became progressively younger toward the Yellowstone area of northwestern Wyoming where large eruptions have occurred as recently as 70,000 years ago. A batholith, in part still molten, underlies Yellowstone and igneous activity will probably recur there in the future. The linear Snake River–Yellowstone trend may have been formed by a deep mantle hot spot, similar to that believed to have formed the Hawaiian Islands, as the North American plate drifted slowly westward over it.

In the Miocene the giant Sierra Nevada block formed a highland of low-to-moderate relief, rising to an elevation of perhaps 1000 meters. Uplift in the Pliocene raised the Sierra to its present elevation and Pleistocene glaciation created the rugged topography that is so spectacular today. Fluvial and deltaic deposition west of the Sierra broadened as the shelf beyond the coastal plain prograded. The site of the present San Joaquin–Sacramento Valley had largely

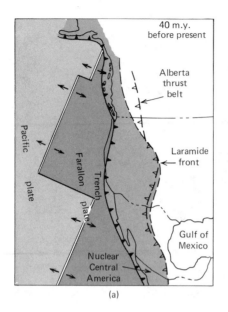

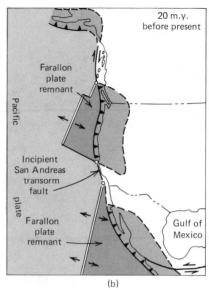

(a)   (b)

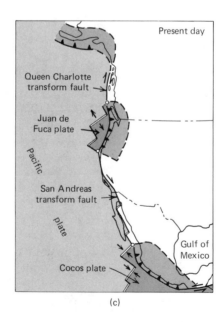

(c)

FIG. 6-7 In the late Eocene (a) the spreading center between the Pacific and Farallon plates was still seaward of the trench. In the Oligocene the spreading center reached the trench and by Miocene (b) the San Andreas Fault was beginning to take up the relative motion between the Pacific and North American plates. (c) Present plate configurations on the west coast of North America. The Cocos and Juan de Fuca plates are all that remain of the Farallon plate. The shaded areas on the continent east of the trench are areas of arc-related volcanism. (Dickinson, 1979)

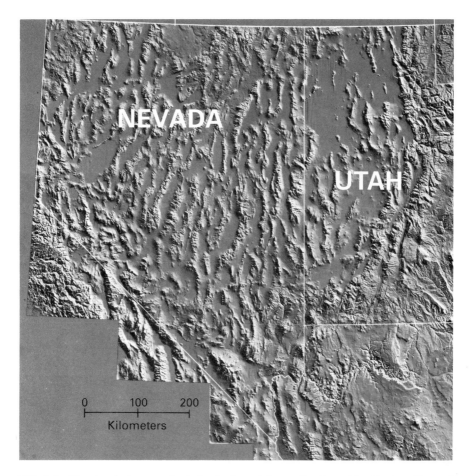

FIG. 6-8 Simulated photograph of the Basin and Range Province. The block uplifts and depressions developed in the Miocene. (Courtesy of the U.S. Geological Survey, Flagstaff, Arizona)

been filled and was largely above sea level by the end of the Pliocene. Deformation associated with the lengthening San Andreas Fault system uplifted the western part of this area, creating the western margin of the valley. Major uplift of the coast ranges occurred in the Pliocene and Pleistocene and very young raised marine terraces show that the uplift is continuing today.

CENOZOIC ENVIRONMENTS

The Deep Sea

Calcareous ooze, which consists of minute skeletons of coccoliths and planktonic foraminifera, covers almost 50 percent of the floor of the deep ocean basins and is by far the most widespread sediment on earth. The other

FIG. 6-9 Cenozoic volcanic rocks of the Pacific northwest showing location of the Columbia River Plateau and Snake River Plain. (Gilluly, 1963)

50 percent of the deep ocean floor is covered largely by clays and siliceous ooze. The distribution of calcium carbonate in sediments of the oceans far from land is largely governed by depth of water. Calcium carbonate dissolves much more rapidly in the very deep ocean than in shallow water. So the deepest parts of the oceans today are accumulating virtually no calcium carbonate. The boundary below which calcium carbonate is absent is surprisingly sharp and is known as the *calcite compensation depth*, commonly referred to as the "CCD." The CCD corresponds to the depth at which the rate of dissolution of calcium carbonate is equal to the rate at which it is supplied from the plankton in the water column above. At depths greater than the CCD, the rate of dissolution exceeds the rate of supply; at depths shallower than the CCD, dissolution is less than supply. The CCD has been compared to a "snow line" on land above which snow-capped peaks exist. Like snow lines on land, the CCD is not at the same elevation everywhere. It lies between 4000 and 5500 meters in most of the world's oceans but is shallower at very high latitudes. Figure 6-10 shows the depth of the CCD measured below sea level in the world's oceans.

Exactly what controls the depth of the CCD is not clear. We do know that the rate of dissolution of calcium carbonate increases sharply in the water a short distance above the CCD and that this is the immediate controlling factor. The solubility of calcium carbonate is increased by an increase in the quantity of dissolved carbon dioxide. Cold water has a greater capacity for carbon dioxide than warm water and the depth of the cold "bottom water," which forms a deep bottom layer in all oceans, strongly influences the depth of

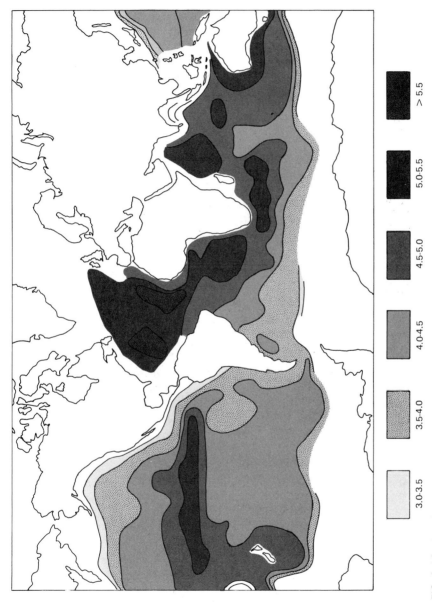

FIG. 6-10 Depth of the calcite compensation depth (CCD) in the world's oceans (in kilometers). (After Berger and Winterer, 1974)

3.0-3.5   3.5-4.0   4.0-4.5   4.5-5.0   5.0-5.5   > 5.5

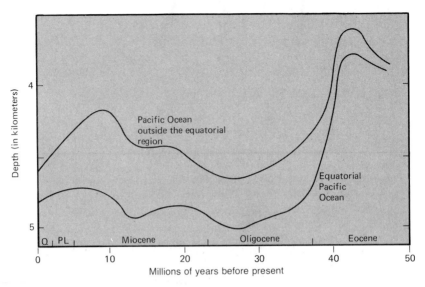

FIG. 6-11 Variation of the CCD during the last 50 million years for the equatorial Pacific and for the Pacific outside the equatorial region. (Van Andel, Heath and Moore, 1975)

the CCD. Where the cold bottom water actually originates, for example, around Antarctica, the CCD is comparatively shallow as Fig. 6-10 shows. The Atlantic Ocean receives the bulk of the world's river runoff (about 70 percent), although in area it constitutes only about 35 percent of the world ocean. As a result, the Atlantic receives the major portion of the calcium carbonate supplied to the ocean by weathering. This tends to depress the CCD in the Atlantic.

Organic productivity of the ocean water in a given area also influences the depth of the CCD. Ocean water contains more carbon dioxide in areas of high productivity than in areas of low productivity because CO_2 is liberated as particulate organic matter oxidizes during its descent from the photic zone through the water column. This fact probably explains the comparatively shallow CCD in the North Pacific, where productivity and carbon dioxide content are relatively high and the comparatively deep CCD in the North Atlantic, where productivity and carbon dioxide content are much lower. In the equatorial Pacific region, where productivity is much greater than in the North Pacific, however, the CCD is considerably deeper. The explanation here appears to be that the extremely high rate of carbonate production and the solution-resistant nature of the equatorial fauna and flora outweigh the effects of increased carbon dioxide. On most continental shelves and slopes the sediments do not contain high percentages of calcium carbonate simply because of the masking effect of large quantities of terrigenous material.

Just as the CCD is not at a constant depth throughout today's oceans, neither has it remained constant with time in any given area. Although evidence is sparse, we infer that it must have been very shallow in the Paleozoic and early Mesozoic before carbonate-secreting plankton evolved. It apparently was also very shallow at the end of the Cretaceous Period. Almost everywhere the Cretaceous–Cenozoic boundary in carbonate sedimentary sequences is represented only by a surface of solution, which seems attributable to a very shallow CCD.

The CCD fluctuated widely during the Cenozoic. The most dramatic excursion occurred during the earliest Oligocene (Fig. 6-11). The late Eocene CCD was relatively shallow (about 3300 meters in the Pacific). It dropped precipitously in the early Oligocene and reached a maximum depth of approximately 5100 meters in the equatorial Pacific by late Oligocene. The Oligocene drop in the CCD was not as great at midlatitudes as at the equator but was still significant (Fig. 6-11).

FIG. 6-12 Relationship of the Green River lake basins to the other Rocky Mountain basins that contain Cenozoic deposits. These basins were produced by block faulting during the Laramide Orogeny in latest Cretaceous and earliest Cenozoic time. (Stokes, 1973)

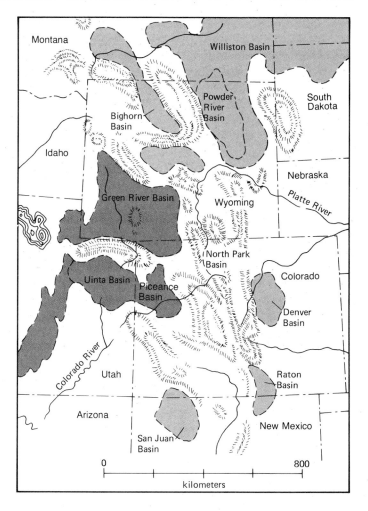

Several tectonic events in the Miocene produced the climatic regime and continental configuration that govern modern oceanographic circulation. The Antarctic ice cap reached its maximum development and was accompanied by the widening of the deep-water passage between Antarctica and Australia. Low-latitude oceanic circulation from the Pacific to the Atlantic was closed off as Africa collided with southern Europe and the Australian plate moved northward into the southeast Asian region. At the same time, the CCD rose several hundred meters to approximately its present position. Fluctuations in the CCD since the Pliocene have been on the order of several hundred meters and are primarily related to glacial-interglacial cycles.

Lacustrine Oil Shales

The world's largest single petroleum reserve, the Green River Formation, formed during Eocene time in large inland lakes in Colorado, Utah, and Wyoming. The petroleum in the Green River Formation is in the form of "oil shale," which is black or dark brown, laminated sedimentary rock composed chiefly of very fine grains of dolomite, calcite, and organic material. These rocks yield oil on heating, the richest as much as 75 gallons per ton of shale. The Green River was deposited in three major depositional basins (Fig. 6-12) and the richest and thickest deposits of oil shale occur in the central portions of the basins where they are interbedded with evaporite deposits consisting chiefly of halite and sodium carbonate minerals, such as trona (hydrous sodium carbonate). These deposits indicate that the lakes themselves must have been extremely saline, much like Great Salt Lake today. The Green River Formation covers 100,000 square kilometers and attains a thickness of about 1000 meters (Fig. 6-13); as a result, it contains an enormous quantity of oil. In the Piceance (Péa-ance) Basin of Colorado alone those shales that assay 15 or more gallons of oil per ton contain a total of 900 billion barrels of oil. In comparison, the total reserves and estimated undiscovered resources of conventional petroleum remaining in the United States are estimated to be 140 billion barrels. One day the Green River shales may be exploited for their petroleum on a large scale.

The Green River lakes were formed by block faulting that produced closed basins bounded by mountains. The mountains trapped the precipitation and provided water that flowed into the arid basins. The early lakes were probably quite fresh. They were populated by fish and a number of freshwater invertebrates, including snails and clams. Later the climate became more arid and the salts, as well as the highest-grade oil shales, were produced. These deposits are unusual; there simply are not many high-grade oil shales like the Green River Formation, which suggests, in turn, that the environment of deposition was also unusual. As a result, the nature of the Green River environment has stirred considerable interest and two different models have been proposed.

Some geologists believe that the oil shale and interbedded evaporites in

FIG. 6-13 The Green River Formation makes up the Roan Cliffs in northwestern Colorado. The top of the plateau is about 1100 meters above the Colorado River at the base of the cliffs.

the Green River may have been produced in a series of playa lakes that alternately expanded widely and contracted to near dryness, exposing a wide expanse of mud flats that were fringed by alluvial fans. During dry periods when the lakes shrank, evaporite minerals precipitated from dense brines and calcite and dolomite precipitated on the exposed mud flats. Oil shale formed during wet periods when the lakes expanded and the waters became fresh enough to produce rich organic muck consisting of bottom-dwelling algae and fungi. The varved appearance of the oil shale was produced by seasonal floods that washed tiny carbonate particles into the lakes where they settled out and temporarily covered the organic muck accumulating on the bottom.

Most geologists who have worked on the Green River believe instead that it was deposited in a permanent, fairly deep lake. They point out that individual thin beds of salt and oil shale can be traced for tens of kilometers. This factor suggests a uniform, widespread, low-energy environment that is easy to visualize in a large permanent lake but difficult to visualize for a playa. In addition, individual deltas built by inflowing streams at the lake margins achieved thicknesses of a few hundred meters and, to some geologists who have studied them, they indicate that the water was a few hundred meters deep. Finally, successive salt beds show a progressive concentration of bromine. This trend suggests a gradual change in water chemistry that might be expected in a permanent lake but not in a periodically dry playa. Very late in their history the Green River lakes did indeed become shallow and finally ceased to exist when the basins became filled with sediment.

Late Cenozoic Glaciations

Today glaciers cover about 10 percent of the earth's land surface; these are located in Antarctica, Greenland, Iceland, and in scattered mountain ranges throughout the world. During their maximum extent in the Pleistocene Epoch, which began more than 2 million years ago, glaciers covered more than 30 percent of the earth's land surface or about 44 million square kilometers (Fig. 6-14). In Europe an ice sheet spread southward from Scandinavia across the Baltic Sea into Germany and Poland; the Alps and the British Isles supported their own ice caps. Continental glaciers extended throughout the northern plains of Russia and large sections of Siberia. They covered the Kamchatka Peninsula and the high mountains and plateaus of Central Asia. The northern part of North America was covered; the southernmost extent of the continental ice sheets corresponds closely to the courses of the Missouri and Ohio rivers. The high mountains of the western United States supported small ice caps and valley glaciers. In the southern hemisphere Antarctica was covered by

FIG. 6-14 Extent of glaciation in the northern hemisphere during Pleistocene glacial maxima.

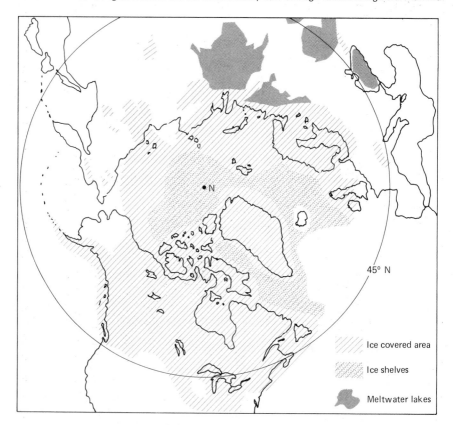

ice, as it is now. (In fact, ice has covered Antarctica thoughout most of the Cenozoic Era). New Zealand, Tasmania, and southern South America were all heavily glaciated. Even in very low latitudes high mountains were glaciated, as in Hawaii and New Guinea.

Glaciers are powerful agents of erosion whose effects are much different from those of running water. As glaciers first form in an area, the streams of ice follow preexisting stream valleys, but as the ice sheets coalesce and thicken, the configuration of the underlying land surface exerts less and less effect on their flow. Fully formed continental glaciers may be more than 3000 meters thick and their direction of flow is determined by the slope of their upper surface, not by slope or irregularities of the rock surface beneath them. Thus the area that receives the greatest net accumulation of snow eventually becomes the central point of radiation for the entire sheet. Ice sheets with their load of rock fragments push down on the land surface with tremendous force. The prolonged advance of a thick ice sheet levels and smooths the underlying surface as the rocks carried along at the bottom of the ice are dragged across the buried landscape. Glaciated surfaces show numerous elongate depressions, striations, and scratches that indicate the direction in which the ice moved (see Fig. 4-25). Beneath ice sheets local areas of hard rock may be left in relief and softer areas may be deeply excavated, but on the whole relief becomes considerably smoothed and subdued. The material thus removed is deposited far downstream by the ice. Boulders the size of a house may be carried hundreds of kilometers.

At the fringes of ice sheets the meltwater produces large streams that carry great quantities of gravel and sand from the glacier as *outwash*. On the Columbia River in Canada, for example, immense terraces of coarse sand and gravel more than 30 meters high attest to huge volumes of meltwater in the recent geologic past. In many areas of northern Europe and North America the advancing glaciers themselves greatly modified drainage patterns and produced ephemeral lakes. Preexisting stream valleys that were gouged by the ice commonly became large depressions. Those that subsequently filled with freshwater became finger lakes (the elongate lakes of upper New York state are examples) and those that filled with the sea became fjords. The Great Lakes basins had a similar origin and underwent a complex history of excavation during successive advances of the ice. Large lakes formed in the northern Great Plains and northern Rocky Mountains. Generally they were dammed by glacial deposits and, in some cases, by the margins of the ice sheet itself and readily filled with meltwater. Occasionally dams made of glacial material broke, causing catastrophic flooding in downstream areas. In western Montana, for example, a large basin filled with water to form glacial Lake Missoula (Fig. 6-15). About 18,000 years ago its dam of moraine material and ice abruptly broke and a wall of water rushed across the Idaho panhandle and eastern Washington with incredible speed. This catastrophic flood scoured channels and deposited immense gravel bars over a large part of the Columbia

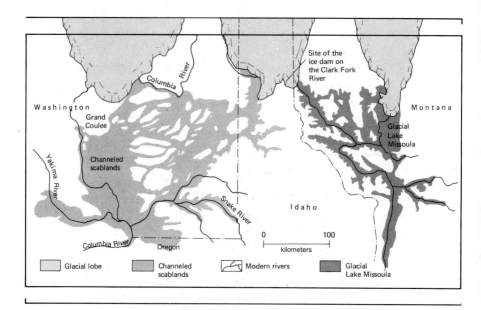

FIG. 6-15 The channeled scablands of eastern Washington were produced about 22,000 years ago by a catastrophic flood caused by the breakup of the ice dam that contained glacial Lake Missoula. (Baker, 1973)

Plateau, creating a unique landscape that has become known as "the channeled scablands."

Glacially induced wet climates also produced huge, deep lakes in the now-arid basin and range region of the western United States, far distant from the ice-sheet margin. The two largest were Lake Bonneville, of which Great Salt Lake is a shrunken remnant, and Lake Lahontan in western Nevada (Fig. 6-16).

The water that was locked up in the ice of Pleistocene glaciers originally came from the oceans. It was evaporated, transported landward by winds, precipitated as snow, and compacted into ice. At their maximum extent, the Pleistocene ice sheets had a volume of about 76.7 million cubic kilometers. As a result of the conversion of seawater to ice, sea level was about 120 meters lower than it is now. At present, about 26.25 million cubic kilometers of ice still remain on the continents. For the last several thousand years, glaciers have been melting and sea level has been rising. Modern records confirm that the trend continues. If all the glaciers were to melt, the water produced would raise sea level an additional 65 meters. A rise of this magnitude would greatly change the outline of the earth's land areas and would submerge most large cities of the world. In the United States Long Island, most of New Jersey, the Delaware–Maryland–Virginia peninsula, and most of Florida would be flooded.

The crust of the earth sinks gently when it accumulates a sufficiently heavy burden, whether of sediment, water, or ice, and it rises gently when such loads are removed. The overloaded area actually sinks into a layer of higher

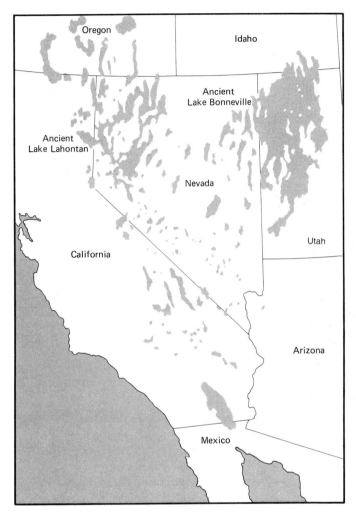

FIG. 6-16 About 15,000 to 20,000 years ago a much more moist climate produced some huge, deep lakes in the now-arid Basin and Range province of the western United States.

density at depth. This is the earth's way of buoying up heavy loads. The condition of balance thus achieved is called *isostasy*.

The concept of isostasy helps to explain the behavior of areas that have supported thick ice caps. The gradual accumulation of a great thickness of ice depresses the earth's crust regionally until a position of equilibrium is reached. At this point, the actual land surface is an ice-covered area that once was above sea level but may now be below it. In the central part of Greenland, for instance, the ice sheet is more than 3000 meters thick and its base is about 350 meters below sea level. If the ice were taken away, the area would temporarily become a shallow sea. A few thousand years after the ice was removed, however, the crust would rise to a new position of equilibrium. A good example of an ice-depressed area that is still in the process of rebounding is Hudson Bay. This shallow epicontinental sea occupies the site once covered by the thickest portion of the ice sheet in North America. Raised marine deposits show that about 240 meters of rebound has occurred since the ice melted. The area continues to emerge much faster than the worldwide rise in sea level (Fig. 6-17). The rise of land around the Baltic Sea is even better known because the region is more densely populated and observations have been made over a longer period. Tide gauge records and other evidence of former water levels show that the land is still emerging. Since the ice disappeared the region has risen at least 76 meters — an average of about one centimeter per year.

Pleistocene Paleoceanography

The effects of Pleistocene climatic changes were not felt solely on land. Studies of pelagic sediment cored from the ocean bottom show that the distributions of oceanic water masses fluctuated widely as Pleistocene glaciers

FIG. 6-17   Present rate of uplift in meters per 100 years in northeastern North America. The southern Hudson Bay area already has rebounded about 240 meters since the ice sheet melted. (Andrews, 1970)

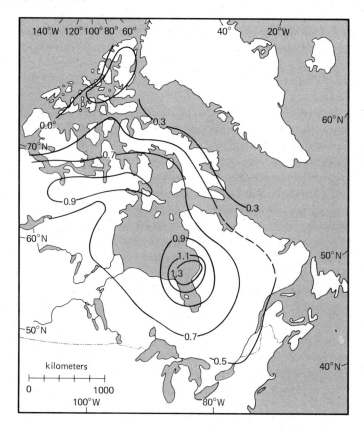

waxed and waned. Moreover, although there were numerous periods of reduced ice volume, climatic conditions were more glacial during most of the Pleistocene than they are today.

Today the distributions of many plankton species are essentially coincident with the surface water masses of the oceans. When calcium carbonate-secreting plankton die, their abandoned shells settle to the bottom, carrying with them a record of the species living in the waters above. For the most part, the species alive at present are the same ones that lived during the Pleistocene. Paleoceanographers have used the records of past distributions of plankton species to discover how the circulation and distribution of water masses change from glacial to interglacial. It has even been possible to make some quantitative estimates of glacial-age sea surface temperatures and degree of seasonal contrast (Fig. 6-18). Equations that relate species abundances in core tops to modern sea surface temperatures have been developed. Pleistocene species abundances are then fed into such *transfer functions* and an estimate of Pleistocene paleotemperatures is calculated. By plotting many such estimates on maps of the oceans, paleoceanographers have developed a picture of how the currents and water masses were distributed during glacial times.

Today the North Atlantic polar front, the southern boundary of the polar water mass, extends as far south as Newfoundland (~45°N) on the western side of the North Atlantic Ocean [Fig. 6-18(a)]. Eastward it is found progressively farther north and it lies between 60° and 65°N from Iceland to Scandinavia. The subpolar and transition water masses occupy most of the area between the polar front and 40°N. But 18,000 years ago, during the last glacial maximum, the distribution of these water masses was quite different [Fig. 6-18(b) and (c)]. The subpolar and transition water masses were squeezed into a narrow band between the subtropical water mass, which had changed very little in its extent [Fig. 6-18(d)], and the polar front, which lay as much as 20° south of its present position, along a nearly east–west line at about 42°N.

At present, the Gulf Stream carries warm water northeastward from off the middle Atlantic coast of the United States toward northern Europe and the British Isles. This warm water makes the climate in northern Europe and Britain much milder than it is at similar latitudes in northern Canada on the west side of the Atlantic Ocean. During glacial times the Gulf Stream turned due east from the middle Atlantic coast of the United States and much of northern Europe was covered with glacial ice. In the Antarctic Ocean the polar front was also shifted much farther equatorward than it is today, but the southward displacement of polar waters was less dramatic in the North Pacific. In other respects, glacial changes in water mass distributions in the Pacific Ocean were similar to the Altantic.

In present-day equatorial regions the prevailng trade winds cause oceanic surface waters to be transported away from the equator—to the north in the northern hemisphere and to the south in the southern hemisphere. Water rises from deeper layers to replace the transported surface waters, creating an equatorial zone of upwelling. Today this upwelling is generally most vigorous on

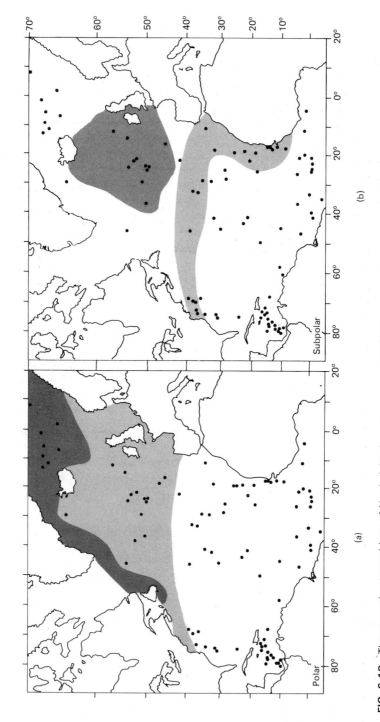

FIG. 6-18 The approximate positions of North Atlantic water masses 18,000 years ago (light) and today (dark) inferred from assemblages of planktonic foraminifera. The polar front moved far to the south during the last glacial age (a). The tropical (e) and subtropical (d) distributions changed very little. (McIntyre and others, 1976)

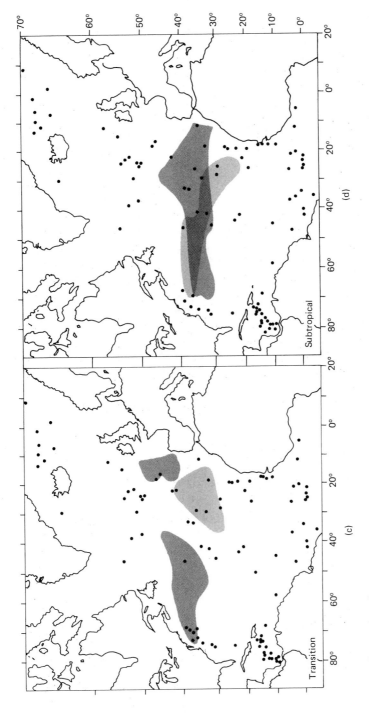

FIG. 6-18 continued.

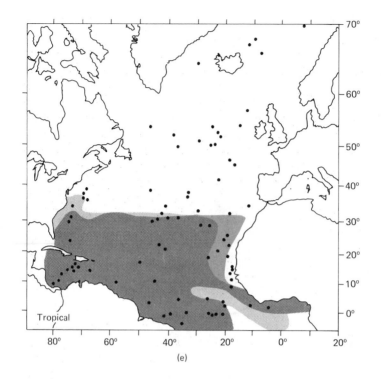

FIG. 6-18 continued.

the eastern sides of the oceans but during glacial times strong thermal gradients enhanced atmospheric circulation, which caused oceanic surface currents to flow faster than they do now. As a result, during glacial times a belt of vigorous upwelling extended across most of the world's oceans at the equator.

When cold waters from high latitudes meet warmer waters at oceanographic fronts, the colder waters generally sink beneath the warmer, less dense waters from lower latitudes. These cold waters from high latitudes form the bottom waters in the world oceans. As polar fronts move equatorward during glacial events, the distributions of bottom waters also shift. Figure 6-19 shows how the distributions of assemblages of deep-living benthic foraminifera changed in the North Atlantic as the polar front moved far to the south.

Oxygen Isotope Stratigraphy and Paleotemperatures

The distribution of oceanic water masses during Pleistocene glacial and interglacial episodes can be reconstructed reliably because the plankton species that lived in that time still exist in the oceans today. Having the same fauna to

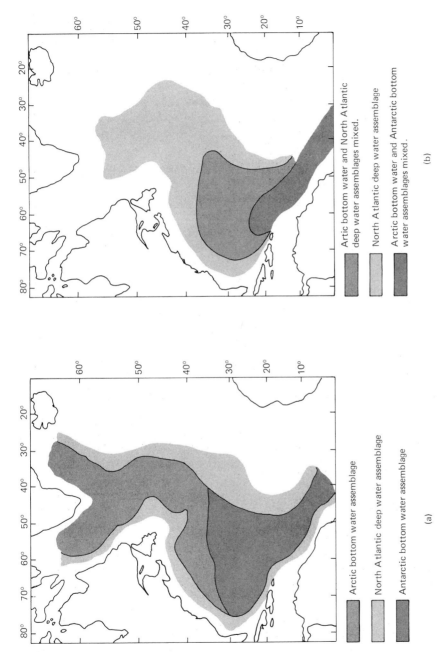

FIG. 6-19 Distributions of benthic foraminiferal assemblages in the North Atlantic today (a) and 17,000 years ago (b) during the last glacial maximum. (Schnitker, 1974)

(a)

Arctic bottom water assemblage

North Atlantic deep water assemblage

Antarctic bottom water assemblage

(b)

Artic bottom water and North Atlantic deep water assemblages mixed.

North Atlantic deep water assemblage

Arctic bottom water and Antarctic bottom water assemblages mixed.

work with back through time makes paleoecologic interpretations relatively easy. Yet it makes faunal correlations difficult because few evolutionary events can be used to date Pleistocene sediments. Magnetic polarity reversals and carbon-14 dates can be used to relate Pleistocene sediments to the absolute time scale, but these methods do not provide a sufficiently detailed stratigraphic framework. Polarity reversals are too widely spaced in time; carbon-14 dating is useful only for the last 40,000 years and requires relatively large quantities of material for analysis.

The development of oxygen isotope stratigraphy has provided an important new tool for Pleistocene correlation. When a foraminifer extracts calcium carbonate from seawater to build its skeleton, it effectively samples the relative proportions of the oxygen isotopes (oxygen-16 and oxygen-18) in the dissolved carbonate ions. These carbonate skeletons thus preserve a record of the isotopic composition of the dissolved ions in the environment in which they grew.

The ratio of oxygen-16 and oxygen-18 in calcium carbonate skeletons depends on several factors. When the temperature is low, carbonate ions dissolved in seawater have a greater proportion of oxygen-18 than when the temperature is high. Thus oxygen isotope ratios in calcium carbonate skeletons fluctuate with cold and warm episodes (corresponding to glacials and interglacials) and they can be used, in some circumstances, to estimate paleotemperatures. Salinity also affects the proportions of oxygen-16 and oxygen-18 in seawater and hence in the dissolved carbonate ions. Increased salinity results from evaporation and water molecules containing oxygen-16 evaporate more readily than those containing oxygen-18. Thus high-salinity water contains a higher proportion of oxygen-18 than low-salinity water. Rain and snow, which derive directly from evaporation, are enriched in oxygen-16 compared to seawater. When precipitation is tied up on the land in the form of glacial ice, the oxygen-16-enriched water is not returned to the ocean so that the proportion of oxygen-18 in the ocean—and hence in foraminiferal skeletons—increases. This is known as the glacial effect. Because glacial episodes are times of decreased temperatures, the glacial effect and the cooling effect are additive. The glacial effect is at least as important as the temperature effect in producing oxygen-18 increases in foraminiferal skeletons formed during glacial episodes.

The glacial-interglacial fluctuations in oxygen isotope ratios for the last 700,000 years are shown in Fig. 6-20. On the right is the standard oxygen isotope curve based on cores from the Caribbean region, with isotope stages numbered, and on the left is a curve from a core taken in the Western Pacific Ocean, with isotope stages also numbered. The correlation is apparent. For deep-sea sediments, oxygen isotope stratigraphy makes possible worldwide Pleistocene correlations with a precision of a few thousand years.

Figure 6-21 shows a generalized curve of ocean surface temperatures (based on planktonic foraminifera) and ocean bottom temperatures (based on benthic foraminifera) inferred from oxygen isotope ratios for the latest Cretaceous and all the Cenozoic. For samples younger than the middle Miocene,

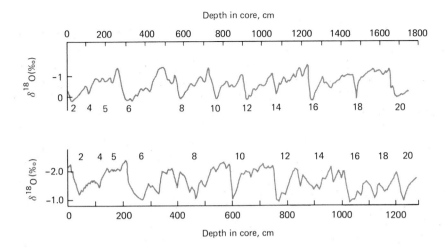

FIG. 6-20 Oxygen isotope stratigraphy for two deep-sea cores, V28-238 in the western Pacific (Shackleton and Opdyke, 1973) and P6408-9 in the Caribbean (Emiliani, 1978). The correlation is apparent. Expressing the isotopic ratios in terms of "delta O-18, parts per thousand" is conventional, as is the numbering of the peaks and valleys on the curves. The time represented is slightly more than 700,000 years.

when significant quantities of ice began to accumulate on land, the glacial effect overshadows the temperature effect and corrections must be applied in order to estimate oceanic paleotemperatures. For the early part of the Cenozoic and for the Cretaceous, paleotemperature estimates are calculated by assuming that, prior to the middle Miocene, the earth was free of continental

FIG. 6-21 Generalized sea surface temperature curves for the last 70 million years based on oxygen-18 analyses. (Douglas and Savin, 1976)

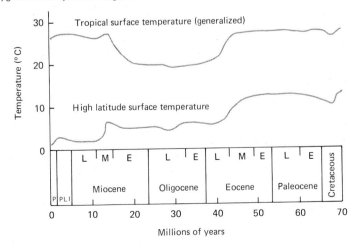

glaciers so that oceanic oxygen isotope ratios showed no significant glacial effect. A major feature of Fig. 6-21 is the progressive divergence in oxygen isotope ratios between surface waters and bottom waters throughout the Cenozoic. It reflects a cooling of high-latitude waters (the source of bottom water) and hence the development of an increased latitudinal thermal gradient. Both atmospheric and oceanic circulation are driven by temperature gradients and so one effect of the increasing gradients must have been increasingly vigorous atmospheric and oceanic circulation—and a more productive ocean as a result.

A puzzling feature of the upper curve in Fig. 6-21 is the comparatively cool tropical sea surface temperature (about 4 to 8° lower than the present) during the middle part of the Cenozoic, from about 17 to 40 million years ago. Temperature estimates from faunas and theoretical considerations suggest that these tropical sea surface temperatures are unreasonably low; that is, they are not real. Even during the last glacial maximum, tropical sea surface temperatures did not decline as much as 4° but instead remained about as warm as they are now. Some geologists believe that the puzzling trough in the oxygen isotope sea surface curve during the middle Cenozoic is not a temperature but a glacial effect. They theorize that although little evidence for continental glaciation of this age has yet been found, a great deal of ice was piling up somewhere on the land at least as early as the Eocene.

The Cause of Ice Ages

What causes an ice age? There are two basic requirements for the formation of a glacier. The first is sufficient precipitation: an arid region could never support a glacier no matter how cold it got. Secondly, it must be cold: summers have to be short and cool enough to prevent complete melting of winter snow.

Numerous theories have been proposed to explain the repeated glaciations of the late Cenozoic. An adequate explanation must account for two items: the initial onset of cold climates and the cyclical repetition of the glaciations. Many geologists feel, too, that an acceptable theory should be applicable to prior ice ages.

The late Cenozoic glaciations are generally presumed to have resulted from a gradual cooling of the entire earth, which began during the Eocene (see Fig. 6-21). This gradual climatic change may have been aided by the worldwide uplift of continents and the resulting withdrawal of the seas from the continents, for widespread epicontinental seas tend to promote equable climates. Plate tectonics may have played a significant role in the cooling of the earth during the Cenozoic. Throughout the Paleozoic and during most of the Mesozoic the North Pole was located in open ocean where oceanic currents could easily transport heat to it. Late Mesozoic and early Cenozoic plate motions formed the isolated Arctic Basin, cutting off free interchange between the waters of the Pacific Ocean and the polar region.

In the southern hemisphere the South Pole was located in the open ocean throughout most of the Mesozoic. By late Mesozoic the Antarctic continent had reached the South Pole and by early Cenozoic it had become centered over the pole. This factor may have accelerated cooling of the southern hemisphere. Continental poles are colder than oceanic ones because summer warmth is not stored by the land, as it is by water, to ameliorate winter temperatures. The Antarctic circumpolar circulation that probably evolved in the early Cenozoic aided the cooling because this circulation prevents warm ocean currents from reaching the Antarctic continental margin.

The closing of the Isthmus of Panama during the Pliocene may have been an important factor in triggering the onset of a glacial climate. This event strengthened the Gulf Stream and caused an increase in precipitation in high northern latitudes, thereby contributing to the accumulation of glacial ice.

What were the repetitive conditions that produced the cyclical glaciations during the Pleistocene? Possibly the cyclical glaciations were triggered by changes in the amount of solar radiation (insolation) received at various latitudes for different times of the year. Cyclic changes in the shape of the earth's orbit and in the inclination of the earth's axis are well known. Using these cycles, the variation in summertime insolation for the northern hemisphere latitudes can be calculated. When plotted against the time scale for the Pleistocene, the calculated insolation changes correlate well with paleotemperatures inferred from oxygen isotope studies (Fig. 6-22). The lag of the temperature minima following insolation declines suggests that astronomical

FIG. 6-22 Cycles of summertime insolation in the northern hemisphere for the last 350,000 years are similar to those of Atlantic Ocean surface water temperatures. (Broecker, 1966)

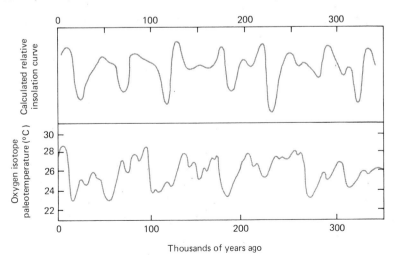

Thousands of years ago

variations may trigger long-term temperature declines. The correlation shown in Fig. 6-22 suggests that slight variations in insolation may well have been responsible for important climatic changes once glaciations began. These astronomical effects did not begin in the late Cenozoic but presumably have operated throughout geologic history; yet major glaciations have been widely spaced in geologic time. This fact suggests that the astronomical effect is small and that it leaves no mark in the historical record except when other factors have operated to reduce the average temperature at high latitudes sufficiently so that slight changes sway the balance between expansion and contraction of glaciers.

Variations in the carbon dioxide content of the atmosphere have been suggested as a possible cause of cyclic glaciations. Increases in the CO_2 level might result at first in warmer and then in colder worldwide climates. Nearly half the incoming solar radiation consists of medium-wavelength visible light to which the atmosphere is transparent. But the reflected radiation from the earth is in the long-wavelength part of the spectrum, which does not easily pass through the atmosphere. Water vapor and CO_2 molecules absorb much of this long-wave energy, which causes them to reradiate heat and thereby warm the atmosphere. Added heat, however, enhances evaporation that would be expected to provide moisture for glaciers and that would increase cloudiness. It has been estimated that an increase of only 2.5 percent in total average annual cloud cover could reduce summer temperatures to levels necessary for a typical glacial episode. Alternatively, *reductions* in atmospheric carbon dioxide could reduce temperatures by allowing more of the earth's radiant energy to escape into space. Unfortunately, we cannot accurately estimate changes in past CO_2 content of either the atmosphere or the oceans. The carbon dioxide budget is very complex and the rate of ocean-atmosphere equilibration is not well known.

Another popular theory proposes that the late Cenozoic glaciations in the northern hemisphere were made possible by the drift of the Arctic Ocean Basin to a polar position. Because the Arctic Ocean has little circulation with the rest of the world's oceans, it is thermally isolated; as a result, the Arctic Ocean and the surrounding northern lands cooled until high-latitude glaciation began in mid-Cenozoic time. Once begun, the increased albedo of perpetually snow-covered lands would accelerate the cooling and the growth of the northern hemisphere ice sheets. According to this theory, the Arctic Ocean eventually froze over and the adjacent North Atlantic and North Pacific oceans cooled substantially. These events would have greatly decreased the moisture available for evaporation and precipitation on nearby lands. Thereafter the continental glaciers could no longer be maintained and began to shrink. The rise in sea level due to glacial melting enhanced the exchange of water between the Atlantic and the Arctic oceans through the Greenland Straits. Eventually, due to the exchange with the Atlantic and to general warming in the northern hemisphere, the Arctic Ocean became ice free.

Increased precipitation from the newly open Arctic could have led to the accumulation of snow on the northern landmasses, just as evaporation from the open Greenland Sea now feeds precipitation on the Greenland ice cap. As glaciers formed anew, temperatures would decrease and the sea level would fall, eventually restricting flow between the Arctic and Atlantic. Ultimately these events would again lead to the freezing over of the Arctic Ocean. The stage was then set for another cycle. According to this theory, the Pleistocene glacial cycles will continue until the North Pole is no longer in the isolated Arctic Basin or until some tectonic event reduces the degree of isolation.

CHANGING LIFE ON LAND

Life in the oceans was already essentially modern in aspect as the Cenozoic Era began; in contrast, many important changes in land-dwelling life have taken place during the last 70 million years. The rapid expansion of flowering plants, which began in the preceding Cretaceous Period, continued throughout Cenozoic time, leading to the extraordinary diversity of land plants that we see around us on the present-day earth. At the same time, land animal life was also changing dramatically.

The Late Cretaceous extinctions of most dominant groups of reptiles set the stage for an explosive evolutionary expansion of mammals early in the Cenozoic. Mammals first evolved from reptilian ancestors during the Triassic Period. These earliest mammals were small, superficially mouselike creatures that had several adaptations not present in most of their reptilian forebears. Among the most useful were mechanisms, such as insulating fur and a more efficient circulatory system, for keeping the body at a constant temperature regardless of the temperature of the surrounding air. This adaptation makes it possible for mammals to remain active in continuously cold climates or during the winter and on cool nights in temperate climates. Most reptiles, in contrast, have no mechanisms for keeping the body warm as the surrounding air temperatures drop and can remain active only in relatively warm air. Because of this adaptive advantage, it is somewhat puzzling that mammals did not expand rapidly after first evolving in early Mesozoic time. Instead they were destined to remain small and relatively uncommon during the Jurassic and Cretaceous periods when dinosaurs dominated the land. Only after the reptile extinctions of Late Cretaceous time did mammals diversify and expand to dominate the land.

Once the mammalian expansion began, it took place rapidly. Early in the Cenozoic Era several groups of large herbivores and carnivores developed; by the middle of the era most of the diversely specialized mammalian groups known today were established (Fig. 6-23). Among them were not only such familiar ground-dwelling herbivores as rodents, horses, camels, pigs, rhinoceroses, and elephants but also carnivorous cat and bearlike forms and such

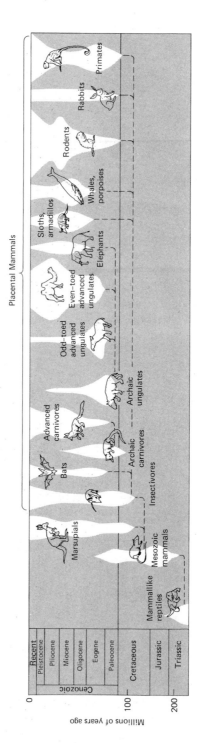

FIG. 6-23 Evolutionary history of the mammals. Dashed lines show the most likely evolutionary relationships among the groups. The width of the vertical bars indicates the approximate abundance of each group.

186

specialized groups as the tree-dwelling primates, flying bats, and ocean-dwelling seals, porpoises, and whales (Fig. 6-24). Indeed, several common mammal groups reached their maximum diversity near the middle of the era and have slowly declined in importance since.

When reptiles first expanded in late Paleozoic and early Mesozoic time, the present continental fragments were still joined into the supercontinent of Pangaea. As a result, early reptiles were able to migrate over much of the land surface and were generally similar everywhere. Similarly, when the first mammals arose in early Mesozoic time, they were widely distributed over the land surface; in contrast, much of the diversification of mammals took place after the mid-Mesozoic breakup of the continents so that intermingling and migration of mammals between continents were often impeded. Consequently, mammalian history has followed a somewhat different course on several continents.

North America, Africa, and Eurasia were interconnected through much of the Cenozoic Era and most of the familiar present-day mammal groups arose and were continuously abundant on these continents. In contrast, South America and Australia, which were isolated from the other continents throughout much of Cenozoic time, each developed its own peculiar mammalian fauna. Mammalian history in South America is complicated by the late Cenozoic linkage with North America, which permitted mixing of the unique South American mammals with more advanced forms from the north. Australia, on the other hand, has been continuously isolated and has retained its

FIG. 6-24 The geologic record of primates. The width of the vertical areas indicates the approximate abundance of each group. The larger modern apes and man are believed to have evolved from tailless, upright Miocene primates called dryopithecines.

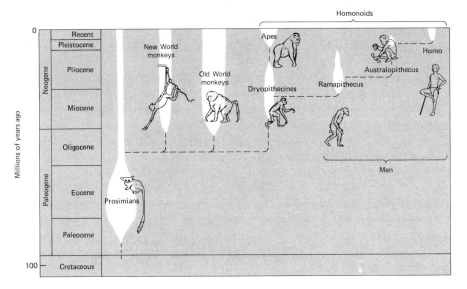

distinctive mammals. All are descendants of primitive marsupial ancestors that differed from other mammals principally in their less advanced mode of reproduction. The Australian expansion of these early marsupials led not only to such distinctive mammals as the kangaroo but also to forms that mimic, in external appearance and habit, such groups as the dogs, cats, and bears that arose on the larger, interconnected continents.

Perhaps of greatest interest is the history of the primates, the mammalian group that includes our own species *Homo sapiens*. The first primates of early Cenozoic time diverged from their less-specialized ancestors by developing adaptations for living in trees rather than on the ground. Among these adaptations were flexible fingers and an opposable thumb for securely grasping tree limbs, plus a shift in the position of the eyes toward the front of the face to afford precise, stereoscopic vision, a prime necessity if a tree dweller is to avoid fatal falls. These two key adaptations are characteristic of all primates, including humans; indeed, they provided the basis for humans' later evolutionary success on the ground, for the grasping hand and precise vision proved ideal for shaping and using tools.

The only primates during early Cenozoic time were several relatively small, primitive types known as *prosimians* ("pre-monkeys"). In mid-Cenozoic time three groups of larger and more advanced primates arose from these prosimian ancestors (Fig. 6-24). One of these groups, the "New World monkeys," includes the monkeys of South America, all of which developed relatively independently on that isolated continent. A second group of "Old World monkeys" developed at the same time in Africa and became widely distributed in both Africa and southern Eurasia. It is, however, the third and final group of advanced primates that arose in mid-Cenozoic time, the *hominids*, that is of greatest interest to us, for it includes not only our own species but also the four surviving groups of apes (chimpanzees, gorillas, orangutans, and gibbons) and several extinct species of apes and man.

The earliest hominoids of mid-Cenozoic time, called *dryopithecines*, were relatively small animals, a little less than a meter high, which differed from their monkey relatives in lacking a tail and in having a more upright posture (Fig. 6-24). Both features were probably adaptations for spending more time on the ground than their mostly tree-dwelling monkey contemporaries did.

The oldest fossil remains that are clearly human are found in sedimentary rocks of late Miocene and early Pliocene age. The exciting discovery of *Australopithecus afarensis* in Pliocene beds of East Africa has thrown new light on the antecedents of humans. The oldest of these finds is well dated radiometrically at 3.6 to 3.8 million years. The teeth, jaws, and skulls of these hominids are primitive and the brain size is small. But footprints preserved in volcanic ash show that these very early hominids walked with a fully upright bipedal gait. Until these discoveries were made, many paleoanthropologists assumed that bipedalism, modern dentition, and increased brain size had evolved together as a system, so to speak, with changes in one feature correlating with and enhancing the impetus for changes in the others. The confirmation of erect bipedalism in such primitive forms requires a new outlook on the evolutionary relationships between these evolving human characteristics.

Slightly younger (2.6 to 3.3 million years) finds in East Africa appear to represent the same species, and skeletal configurations confirm the bipedalism of *Australopithecus afarensis*. These hominids seem to be the ancestors of two diverging evolutionary lines. One line led to a robust form of *Australopithecus* that was probably a vegetarian, for it had large jaws and teeth adapted to grinding. This line became extinct in the early Pleistocene. The other line led to modern humans . The descendants of *A. afarensis* learned to make primitive stone tools and to use fire.

The first *Homo sapiens* appeared about 125,000 years ago and is known as *Neanderthal Man*. These people were sturdily built, with heavy bones that suggest an active and rugged life. They used tools and fire and they buried their dead in graves. Truly modern humans appeared approximately 50,000 years ago, about the time the Neanderthals died out. It is not certain whether modern humans were the direct descendants of the Neanderthals or if they were a second geographical variant or race that simply displaced the Neanderthals ecologically and culturally.

It was during the climatic extremes of the Pleistocene Epoch that the genus *Homo* evolved in Africa or Asia and, in a very short time geologically, dispersed throughout the world. In the past two centuries, humans have altered the earth's surface to an unprecedented degree. The effect of industrial civilization on the land, the atmosphere, and the ocean represents the most rapid environmental disruption caused by any organism in all geologic history. Only through continued study of the earth's crust can men and women learn to understand it better so that they may assess their impact on it and address their prospects for the future.

bibliography

ANDREWS, J. T. 1970. Present and post-glacial rates of uplift for glaciated northern and eastern North America derived from postglacial uplift curves. *Can. Jour. Earth Sciences* 7: 703–715.

BAKER, V. R. 1973. Paleohydrology and sedimentology of Lake Missoula flooding in eastern Washington. *Geol. Soc. America Special Paper* 144: 79 pp.

BARAGER, W. R. A., and MCGLYNN, F. C. 1976. Early Archean basement in the Canadian Shield: A review of the evidence. *Geol. Survey Canada Paper* 76–14.

BECK, M.; COX, A.; and JONES, D. L. 1980. Mesozoic and Cenozoic microplate tectonics of Western North America. *Geology* 8: 454–456.

BERGER, W. H., and WINTERER, E. L. 1974. Plate stratigraphy and the fluctuating carbonate line, in *Pelagic sediments on land and under the sea*, Hsu and Jenkyns, eds. Oxford: International Assoc. of Sedimentologists, Blackwell Scientific Publications, Ltd., Special Publication 1: 11–48.

BROECKER, W. S., 1966. Absolute dating and the astronomical theory of the glaciation. *Science* 151, no. 3708:299-304.

BURKE, K. 1975. Atlantic evaporites formed by evaporation of water spilled from Pacific, Tethyan, and Southern Oceans. *Geology* 3: 613–616.

COOK, H. E., and TAYLOR, M. E. 1977. Comparison of continental slope and shelf environments in the Upper Cambrian and lowermost Ordovician of Nevada, in *Deep-water Carbonate Environments*, Cook and Enos, eds. Tulsa: Soc. Econ. Paleont. Miner., Spec. Publ. 25: 51–81.

DAVIS, G. A. 1974. Pre-Mesozoic history of California and the West: San Joaquin Geol. Soc., *Plate tectonics short course*, pp. 10-1 — 10-6.

DICKINSON, W. R. 1979. Cenozoic Plate Tectonic setting of the Cordilleran Region in the United States, in *Cenozoic Paleogeography of the Western United States*, Armentrout, Cole and TerBest, eds., Los Angeles: Pacific Section, Soc. Econ. Paleont. Miner: 1–13.

DIETZ, R. S. 1972. Geosynclines, mountains, and continent-building. *Scientific American* 226, no. 3: 30–38.

DIETZ, R. S. and HOLDEN, J. C. 1970. Reconstruction of Pangaea: Break-up and dispersion of continents, Permian to Recent. *Jour. Geophysical Research* 75: 4939–4956.

EMILIANI, C. 1978. The cause of the ice ages. *Earth and Planetary Sci. Letters* 37: 349–354.

FRASER, J. A.; HOFFMAN, P. F.; IRVINE, T. N.; and MURSKY, G. 1972. The Bear Province: in Price and Douglas, eds., *Variations in Tectonic Styles in Canada*. Geol. Assoc. Canada, Special Paper 11: 453–503.

GILLULY, J. 1963. The tectonic evolution of the western United States. *Quart. Journ. Geol. Soc. London*, 119: 133–174.

GRIEVE, R. A. F. 1980 Impact bombardment and its role in proto-continental growth on the early earth. *Precambrian Research* 10: 217–247.

HARRIS, A. G.; HARRIS, L. D.; and EPSTEIN, J. B. 1978. Oil and gas data from Paleozoic rocks in the Appalachian Basin; maps for assessing hydrocarbon potential and thermal maturity (conodont color alteration isograds and over-burden isopachs). *U.S. Geol. Surv., Misc. Invest. Ser.*, No. I–917–E.

HOFFMAN, P.; DEWEY, J. F.; and BURKE, J. 1974. Aulacogens and their genetic relation to geosynclines, with a Proterozoic example from Great Slave Lake, Canada, in *Modern and Ancient Geosynclinal Sedimentation*, Dott and Ahaver, eds. Tulsa: Soc. Econ Paleont. Miner., Special Publication 19: 38–55.

HSÜ, K. J. 1972. When the Mediterranean dried up. *Scientific American* 227, no. 6: 26–36.

KARIG, D. E. and others. 1979. Structure and Cenozoic evolution of the Sunda Arc in the central Sumatra region. *Am. Assoc. Petrol. Geologists Memoir* 29: 223–238.

KAUFFMAN, E. G. 1973. Cretaceous assemblages, communities and associations; western interior United States and Caribbean Islands. In *Principles of Benthic Community Analysis*, Ziegler, A. M., ed., Sedimenta IV, Comp. Sed. Lab., Univ. Miami, pp. 12.1–12.27.

KING, P. B. 1948. Geology of the southern Guadalupe Mountains, Texas. *U.S. Geol. Survey, Professional Paper* 215: 183 pp.

KOLB, C. R., and VAN LOPIK, J. R. 1966. Depositional environments of the Mississippi River deltaic plain—southeastern Louisiana, in *Deltas in their geologic framework*, Shirley and Ragsdale, eds., Houston, Texas: Houston Geol. Society, pp. 17–62.

KREBS, W. 1974. Devonian carbonate complexes of Central Europe. In *Reefs in Time and Space*, L. F. Laport, ed., Tulsa: Soc. Econ. Paleont. Miner., Spec. Publ. 18: pp. 155–208.

LAND, L. S. 1973. Holocene meteoric dolomitization of Pleistocene limestone, North Jamaica. *Sedimentology* 20: 411–424.

LAPORTE, L. F. 1968. *Ancient environments*. Englewood Cliffs, N.J.: Prentice-Hall, Inc., 115 pp.

MCINTYRE, A.; KIPP, N. G.; BÉ, A. W. H.; CROWLEY, T.; KELLOG, T.; GARDNER, J. V.; PRELL, W. N.; RUDDIMANN. 1976. Glacial North Atlantic 18,000 years ago: A CLIMAP reconstruction, in *Investigation of Late Quaternary Paleoceanography and Paleoclimatology*, Cline and Hays, eds., U.S. Geol. Soc. America Memoir 145: 43–76.

MCKEE, E. D. and others. 1975. Paleotectonic investigations of the Pennsylvanian System in the United States. *U.S. Geol. Survey, Professional Paper* 853.

MCKENZIE, D. P. 1972. Plate tectonics and sea-floor spreading. *Am. Scientist* 60: 425-435.

MATTHEWS, R. K. 1974. *Dynamic stratigraphy: an introduction to sedimentation and stratigraphy* Englewood Cliffs, N.J.: Prentice-Hall, Inc., 370 pp.

MESOLELLA, K. J. 1978. Paleogeography of some Silurian and Devonian reef trends, Central Appalachian Basin. Am. Assoc. Petrol. Geologists Bull., 62: 1607–1644.

NILSEN, T. H., and MCKEE, E. H. 1979. Paleogene paleogeography of the western United States, in *Cenozoic Paleogeography of the Western United States*, Armentrout, Cole and Ter Best, eds., Los Angeles: Pacific Section, Soc. Econ. Paleont. Miner., pp. 257–276.

POOLE, F. G.; SANDBERG, C. A.; and BOUCOT, A. J. 1977. Silurian and Devonian paleogeography of the western United States, in *Paleozoic Paleogeography of the Western United States*, Stewart, Stevens, and Fritsche, eds., Los Angeles: Pacific Section, Soc. Econ. Paleont. Miner., pp. 39–66.

POOLE, F. G. and SANDBERG, C. A. 1977. Mississippian paleogeography and tectonics of the Western United States, in *Paleozoic Paleogeography of the Western United States*, Stewart, Stevens, and Fritsche, eds., Los Angeles: Pacific Section, Soc. Econ. Paleont. Miner., pp. 67–86.

PRELL, W. L. and HAYS, J. D. 1976. Late Pleistocene faunal and temperature patterns of the Columbia Basin, Caribbean Sea, in *Investigations of Late Quaternary Paleoceanography and Paleoclimatology*, Cline and Hays, eds., Geol. Soc. America Memoir 145: 201–220.

SCHLEE, J. S.; FOLGER, D. W.; DILLON, W. P.; KLIDGORD, K. D.; GROW, J. A., 1979. The continental margins of the Western North Atlantic. *Oceanus* 22, no. 3: 40–47.

SCHNITKER, D. 1974. Western Atlantic abyssal circulation during the past 12,000 years. *Nature* 248: 385–387.

SCOTESE, C.; BAMBACH, R. K.; BARTON, C.; VAN DER VOO, R.; and ZEIGLER, A. 1979. Paleozoic base maps. *Jour. Geology* 87, no. 3: 217–277.

SEARS, S. O., and LUCIA, F. J. 1980. Dolomitization of northern Michigan Niagara Reefs by brine refluxion and freshwater/seawater mixing, in *Concepts and Models of Dolomitization*, Zenger, Dunham, and Ethington, eds., Tulsa, Soc. Econ. Paleont. Miner., Spec. Publ. 28: 215–236.

SEELY, D. R. 1979. The evolution of structural highs bordering major forearc basins, in *Geology of Continental Margins*, Watkins, J. S., ed., Am. Assoc. Petrol. Geologists Memoir 29: 245–260.

SEYFERT, C. K. and SIRKIN, L. A. 1973. *Earth history and plate tectonics, an introduction to historical geology*. New York: Harper & Row, Pubs., 504 pp.

SHACKLETON, N. J. and OPDYKE, N. D. 1973. Oxygen isotope and paleomagnetic stratigraphy of Equatorial Pacific Core V28-238: Oxygen isotope temperatures and ice volumes on a 10^5

and 10^6 year scale. *Quaternary Research* 3: 39–55.

———. 1976. Oxygen isotope and paleomagnetic stratigraphy of core V28-239, late Pliocene to late Pleistocene. *Geol. Soc. America Memoir* 145: 449–464.

SLOSS, L. L. 1963. Sequences in the cratonic interior of North America. *Geol Soc. America Bull.* 74: 93–114.

STEWART, J. H. 1972. Initial deposits in the Cordilleran Geosyncline: Evidence of a late Precambrian (less than 850 m.y.) continental separation. *Geol. Soc. America Bull.* 83: 1345–1360.

SULLIVAN, WALTER. 1974. *Continents in motion: The new earth debate.* New York: McGraw-Hill, 399 pp.

WICKHAM, J.; ROEDER, D.; and BRIGGS, G. 1976. Plate tectonics models for the Ouachita foldbelt. *Geology* 4: 174–176.

WILSON, J. L. 1975. *Carbonate Facies in Geologic History.* New York: Springer-Verlay, 471 pp.

WINDLEY, B. F. 1976. *The evolving continents.* New York: John Wiley and Sons, 385 pp.

index

Relative Durations of Major Geologic Intervals	Era	Period	Epoch	Duration in Millions of Years (Approx.)	Millions of Years Ago (Approx.)	
CENOZOIC		Neogene	Recent	Approx. last 10,000 years		0
			Pleistocene	2	2	
MESOZOIC			Pliocene	3	5	
			Miocene	20	25	50
		Paleogene	Oligocene	13	38	
PALEOZOIC			Eocene	17	55	
	Cenozoic		Paleocene	10	65	
						100
		Cretaceous		79	144	150
		Jurassic		69	213	200
	Mesozoic	Triassic		35	248	250
		Permian		38	286	300
		Pennsylvanian (Carboniferous)		34	320	
		Mississippian (Carboniferous)		40	360	350
		Devonian		48	408	400
		Silurian		30	438	450
		Ordovician		67	505	500
						550
	Paleozoic	Cambrian		85	570	
PRECAMBRIAN	Precambrian			4,030		4,600

Formation of Earth's crust about 4,600 million years ago

Millions of Years